11th International Conference on Mathematics and Engineering Physics (ICMEP-11)

Journal of Physics: Conference Series Volume 2304

Cairo, Egypt
29-31 March 2022

ISBN: 978-1-7138-6031-0
ISSN: 1742-6588

Printed from e-media with permission by:

Curran Associates, Inc.
57 Morehouse Lane
Red Hook, NY 12571

Some format issues inherent in the e-media version may also appear in this print version.

Institute of Physics
Dirac House, Temple Back
Bristol BS1 6BE UK

Phone: 44 1 17 929 7481
Fax: 44 1 17 920 0979

techtracking@iop.org

Additional copies of this publication are available from:

Curran Associates, Inc.
57 Morehouse Lane
Red Hook, NY 12571 USA
Phone: 845-758-0400
Fax: 845-758-2633
Email: curran@proceedings.com
Web: www.proceedings.com

TABLE OF CONTENTS

ENGINEERING MATHEMATICS

ENGINEERING PHYSICS

Arab Republic of Egypt
Ministry of Defense
Military Technical College

The 11th International Conference on Mathematics and Engineering Physics

29-31 March, 2022

(ICMEP-11)

Conference Program, Keynotes, Workshops and Seminars of (ICMEP-11)

Military Technical College
Kobry El-Kobbah,
Cairo, Egypt.
icmep@mtc.edu.eg

11th International Conference
on Mathematics and
Engineering Physics
29-31 March 2022
https://icmep.conferences.ekb.eg/

Journal of Physics: Conference Series **2304** (2022) 011001 doi:10.1088/1742-6596/2304/1/011001

Military Technical College
Kobry El-Kobbah,
Cairo, Egypt.
icmep@mtc.edu.eg

11th International Conference on Mathematics and Engineering Physics
29-31 March 2022
https://icmep.conferences.ekb.eg/

Preface

The Military Technical College, sponsored by the Egyptian Ministry of Defense, is honored to organize the ninth International Conference on Mathematics and Engineering Physics in the period of March 29 - 31, 2022.

The conference provides a valuable opportunity to all the research people in Egypt and all over the world to exchange and update their knowledge in the designated fields of interest

The papers accepted by the scientific committee were presented during the conference period in 4 sessions. The conference proceedings include only the full-text papers, which were presented during the conference and approved by the respective session boards. Accepted papers are arranged in 4 sessions in the following scientific branches:

Engineering Mathematics, this includes:

- **Numerical Analysis,**
- **Mathematical Analysis.**
- **Probability and Statistics.**
- **Fuzzy Mathematics**
- **Optimization**

Engineering Physics, this includes:

- **Applied Optics & Holographic Techniques.**
- **Generation of High-Power Laser & Simulation.**
- **Lasers and their applications in medicine.**
- **Laser–Material or Biological Tissue Interaction.**
- **Nanoparticles Technology.**

We acknowledge all the effort done by the contributors from the universities, research centers, institutes, industry and Egyptian Armed Forces

Journal of Physics: Conference Series **2304** (2022) 011001 doi:10.1088/1742-6596/2304/1/011001

Military Technical College
Kobry El-Kobbah,
Cairo, Egypt.
icmep@mtc.edu.eg

11th International Conference on Mathematics and Engineering Physics
29-31 March 2022
https://icmep.conferences.ekb.eg/

Conference Committee

Journal of Physics: Conference Series **2304** (2022) 011001 doi:10.1088/1742-6596/2304/1/011001

Military Technical College
Kobry El-Kobbah,
Cairo, Egypt.
icmep@mtc.edu.eg

11th International Conference on Mathematics and Engineering Physics
29-31 March 2022
https://icmep.conferences.ekb.eg/

CONFERENCE HIGH COMMITTEE

Maj. Gen. Prof. Dr. Ismail Mohamed Kamal Ismail	**Chairman**
Maj. Gen. Prof. Dr. Moutaz Abo El-Noor	**Vice Chairman**
Maj. Gen (R) Prof. Dr. Ahmed Elsaid Abdalla	**Gen. Rapporteur**
Maj. Gen.(R) Prof. Dr. Gamal Mabrouk Abdelhamed	**Rapporteur**

CONFERENCE STEERING COMMITTEE

Maj. Gen. (R) Prof. Dr.	G. M. Abdel-Hamed
Brig. Gen. Dr.	Ahmed Salah Ismail
Brig. Gen. Dr.	Hany Nasry Zaky
Brig. Gen. Dr.	Osama Mohamed Helal
Brig. Gen. Assoc. Prof. Dr.	Mohamed Abdo Mohamed
Brig. Gen. Dr.	Ahmed Hassan Mohamed
Col. Dr.	Mohamed Darwiesh Mostafa
Col. Dr.	Ahmed Kamal El-sherif
Lt. Col. Dr.	Walid Gomaa Abdelwahed

CONFERENCE ORGANIZING COMMITTEE

Maj. Gen. (R) Prof. Dr.	G. M. Abdel-Hamed
Brig. Gen. Dr.	Ahmed Salah Ismail
Brig. Gen. Dr.	Hany Nasry Zaky
Brig. Gen. Dr.	Osama Mohamed Helal
Brig. Gen. Assoc. Prof. Dr.	Mohamed Abdo Mohamed
Brig. Gen. Dr.	Ahmed Hassan Mohamed
Col. Dr.	Mohamed Darwiesh Mostafa
Col. Dr.	Ahmed Kamal El-sherif
Lt. Col. Dr.	Walid Gomaa Abdelwahed

Military Technical College
Kobry El-Kobbah,
Cairo, Egypt.
icmep@mtc.edu.eg

11th International Conference
on Mathematics and
Engineering Physics
29-31 March 2022
https://icmep.conferences.ekb.eg/

Contributors

Military Technical College
Kobry El-Kobbah,
Cairo, Egypt.
icmep@mtc.edu.eg

11th International Conference on Mathematics and Engineering Physics
29-31 March 2022
https://icmep.conferences.ekb.eg/

Military Technical College, Cairo, Egypt.
Egyptian Atomic Energy Authority
higher institute of engineering and technology 5th settlement
Computer Science Faculty, Akhbar El-Yom Academy
Menoufia University
The British University in EGYPT
Cairo University
Minia University
Al Azhar University
National Center for Radiation Research and Technology (NCRRT)
Benha University
Modern University for Technology and Information
Egyptian Chinese University
Heliopolis University for sustainable development
Baku State University, Azerbaijan
Higher Institute of Engineering and Technology, Menoufia
University of Victoria
NILES, Cairo University

Journal of Physics: Conference Series **2304** (2022) 011001 doi:10.1088/1742-6596/2304/1/011001

Military Technical College
Kobry El-Kobbah,
Cairo, Egypt.
icmep@mtc.edu.eg

11th International Conference
on Mathematics and
Engineering Physics
29-31 March 2022
https://icmep.conferences.ekb.eg/

Conference Program

Journal of Physics: Conference Series **2304** (2022) 011001 doi:10.1088/1742-6596/2304/1/011001

Military Technical College
Kobry El-Kobbah,
Cairo, Egypt.
icmep@mtc.edu.eg

11th International Conference on Mathematics and Engineering Physics
29-31 March 2022
https://icmep.conferences.ekb.eg/

وزارة الدفـــــــاع
الكلية الفنية العسكرية

Program of the 11th International Scientific Conference

Tuesday 29 March 2022

Hall	9:00 – 10:00	10:00 – 12:00	12:00 – 13:30	13:30 – 13:45	13:45 – 15:15
A12	**Opening the Conference** ❖ Commandant of the MTC ❖ Conference General Rapporteur صالة العاشر	**Opening Lecture** صالة العاشر	Building Technology		Transportation , Water Resources & Geo-Technical
A11			Metallic Structures & Project Management		Concrete Structures
A10			Automotive Engineering		Material Science and Processing-I
اللواء عبد النبي					
A9			Engineering Mathematics-I		Engineering Mathematics-II
A14					
A8			Chemical & Radiation Detection and Analysis-I		Chemical & Radiation Detection and Analysis-II
اللواء مدحت					
A7			Control Systems and Automation-I		Aerospace Navigation Engineering-II
اللواء عبد الحميد					

* Workshops, Invited Talks - ICCAE - AMME -ICMEP - ICEE - ICEENG

وزارة الدفـــــــاع
الكلية الفنية العسكرية

Program of the 11th International Scientific Conference

Wednesday 30 March 2022

Hall	8:30 – 10:00	10:00 – 10:15	10:15 – 11:45	11:45 – 12:00	12:00 – 13:30	13:30 – 13:45	13:45 – 15:15
A12	Urban Planning		Architecture Engineering -I		Architecture Engineering -IV		Environmental Engineering -I
A11	Architecture Engineering-II		Architecture Engineering-III				
A10	Mechanical Design / Mechatronics		Material Science and Processing-II		Ballistics		Dynamics and Vibrations-I
اللواء عبد النبي	Manufacture of armored steel*		New Trends of Swarm Robotics*		Advances in sheet metal forming technologies*		
A9	Engineering Physics-I		Engineering Physics-II		Engineering Physics-III		
A14	~~Advanced Topics In Laser and Quantum Physics Applications*~~		Advanced Topics In Laser and Quantum Physics Applications*				
A8	Industrial Chemical Engineering	—	Nanotechnology and Chemical Applications-I		Nanotechnology and Chemical Applications-II		Energetic Materials Technology-I
اللواء مدحت	Nuclear Energy is Safe, Economical, Sustainable, and Friend to Environment*		Nanobiotic Formulations as Promising Advances for Combating Multidrug-Resistant Pathogenic Microbes*		Role of geological methods and techniques in solving*		Behavior of substructure and superstructure Subjected to Blast Loads by using ABAQUS*
A7	Power and Energy		Control Systems and Automation-II		Computer Engineering and Artificial Intelligence-I		Communications and Networks
اللواء عبد الحميد	Polarization Imaging in Computer Vision*		Developing Entrepreneurial Talents for Aerospace of the Future*		Multi-Sensor Fusion for Autonomous Vehicle's Positioning Research, Innovation and Future Vision*		Anticipatory Radio Resource Management for 5G Networks and Beyond*

* Workshops, Invited Talks - ICCAE - AMME -ICMEP - ICEE - ICEENG

Journal of Physics: Conference Series **2304** (2022) 011001 doi:10.1088/1742-6596/2304/1/011001

Military Technical College
Kobry El-Kobbah,
Cairo, Egypt.
icmep@mtc.edu.eg

11th International Conference
on Mathematics and
Engineering Physics
29-31 March 2022
https://icmep.conferences.ekb.eg/

Location of Conference Rooms

Military Technical College
Kobry El-Kobbah,
Cairo, Egypt.
icmep@mtc.edu.eg

11th International Conference on Mathematics and Engineering Physics
29-31 March 2022
https://icmep.conferences.ekb.eg/

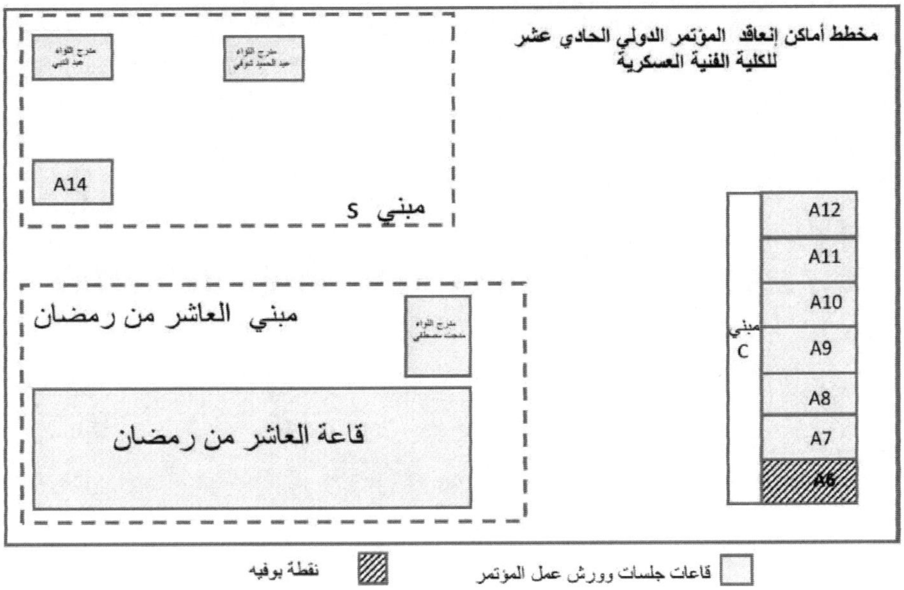

Journal of Physics: Conference Series **2304** (2022) 011001 doi:10.1088/1742-6596/2304/1/011001

Military Technical College
Kobry El-Kobbah,
Cairo, Egypt.
icmep@mtc.edu.eg

11th International Conference on Mathematics and Engineering Physics
29-31 March 2022
https://icmep.conferences.ekb.eg/

Part I: Scientific Sessions

Military Technical College
Kobry El-Kobbah,
Cairo, Egypt.
icmep@mtc.edu.eg

11th International Conference
on Mathematics and
Engineering Physics
29-31 March 2022
https://icmep.conferences.ekb.eg/

Tuesday March 29, 2022 Time: 12:00 - 13:30	Session: EM I Room: A9

Engineering Mathematics I

Chairman: Prof. Dr. Hassan El-Hamouly **M.T.C**
Prof. Dr. Kamal Hassan **BUE**

Maj. Gen. Prof. Dr. Gamal M. Abdel-Hamid **M.T.C**

001_ ICMEP-11	Mathematical Model for Omnidirectional Sensor Network Using Clifford Algebra	Amer Mohamed Mahfouz, Hany Nasry Zaky, Ahmed Salah Ismail, Eiman Taha Abou El Dahab
002_ ICMEP-11	Study and Analysis of Corona-Virus Transfer Dynamics using Enhanced SEIR Epidemic Model with Vaccination Effect	Hazem Magdy Abd El Raouf, Alhaytham M. Aref, Ahmed K. Elsherif, Mohamed E. Khalifail,
003_ ICMEP-11	Applied method of characteristics on 2nd order linear P.D. E	Ahmed Saeed farg, A. M. Abd Elbary, Tarek A. Khalil
004_ ICMEP-11	Analytical solution of diffusion equation under chemical reaction and wet deposition from line Source	Khaled S M Essa, Ahmed M Mosallem and H M Taha

Journal of Physics: Conference Series **2304** (2022) 011001 doi:10.1088/1742-6596/2304/1/011001

Military Technical College
Kobry El-Kobbah,
Cairo, Egypt.
icmep@mtc.edu.eg

11th International Conference
on Mathematics and
Engineering Physics
29-31 March 2022
https://icmep.conferences.ekb.eg/

Tuesday March 29, 2022	**Session: EM II**
Time: 13:45 - 15:15	**Room: A9**

Engineering Mathematics II

Chairman: Prof. Dr. Hassan El-Hamouly M.T.C
Maj. Gen. Prof. Dr. Gamal M. Abdel-Hamid M.T.C

Brig. Dr. Ahmed Salah Ismail M.T.C

005_ ICMEP-11	Generalized thermoelastic in a fiber-reinforced medium with cylindrical cavity under G-N theory	S. E. Khader, A. A. Marrouf and M. Khedr
006_ ICMEP-11	Decoupling and Coupling Approach for the Biharmonic Equation	Kamal Hassan and Mahmoud Farid The British University in EGYPT
007_ ICMEP-11	Multi Chaotic System to Generate Novel S-Box for Image Encryption	Hany Nasry, Azhaar A. Abdallah, Alaa K. Farhan, Hossam E. Ahmed, Wageda I.El Sobky
008_ ICMEP-11	A Numerical Study of Decarbonizing Marine Gas Turbine Emissions Through Ammonia/Hydrogen Fuel Blends	Moataz Medhat, Adel Khalil and Mohamed A Yehia

Journal of Physics: Conference Series **2304** (2022) 011001 doi:10.1088/1742-6596/2304/1/011001

Military Technical College
Kobry El-Kobbah,
Cairo, Egypt.
icmep@mtc.edu.eg

11th International Conference on Mathematics and Engineering Physics
29-31 March 2022
https://icmep.conferences.ekb.eg/

Wednesday March 30, 2022 Time: 8:30 - 10:00	Session: EP I Room: A9

Engineering Physics I

Chairman: Prof. Dr. Amer Mohamed Esmail Kane — **Faculty of Science, Al-Azhar Univ**
Prof. Dr. Ahmed Ashour — **Egyptian Atomic Energy Authority (EAEA)**
Col. Dr. Ahmed Farghl — **M.T.C.**

009_ICMEP-11	Effect of 14.7 MeV Protons on Beryllium Micro Brush / Copper ITER First Wall using SRIM	S. I. Radwan and H. El-Khabeary
010_ ICMEP-11	Simulation of Ozone Concentrations Inside Co 60 Industrial Irradiator Using U.V Optical Cell	Amal El gamel, Ali Noaman
011_ ICMEP-11	Improved treatment of home-made glow discharge ion source	M.M. Abdel Rahman, F.W. Abdel Salam and B. A. Soliman
012_ ICMEP-11	Sr and Rb Mass Interferences under Plasma Conditions	A. M. Rashad, S. A. Abd Elaal and A. M. Abdelhady
013_ ICMEP-11	Spectra of Heavy Quarkonia in a Magnetized-hot Medium in the Framework of Fractional Non-relativistic Quark Model	M. Abu-shady, A. I. Ahmadov, H. M. Fath-Allah and V. H. Badalov

Military Technical College
Kobry El-Kobbah,
Cairo, Egypt.
icmep@mtc.edu.eg

11th International Conference on Mathematics and Engineering Physics
29-31 March 2022
https://icmep.conferences.ekb.eg/

Wednesday March 30, 2022	**Session: EP II**
Time: 10:15 - 11:45	**Room: A9**

Engineering Physics II

Chairman:	**Prof. Dr. Amer Mohamed Esmail Kany**	**Faculty of Science, Al-Azhar Univ**
	Prof. Dr. Ahmed Ashour	**Egyptian Atomic Energy Authority (EAEA)**
	Brig. Dr. Osma Halal	**M.T.C**

014_ICMEP-11	Utilization of INAA Technique for Studying the Negative Impact of Human Activities on Coral Reefs in the Quseir harbor area	Mohamed Y.M. Mohsen, Atef El-Taher, Mohamed Shabib, Nader M. A. Mohamed, H. A. Ashry, Mohamed A.E. Abdel-Rahman
015_ICMEP-11	Effective Hybridization of Biogeography Based Optimization and Simulated Annealing	E. G. Zahran, A. A. Arafa, H.I.Saleh, M.I.Dessouky
016_ ICMEP-11	Application of suggested Three Parameters Collective Rotational Model of Superdeformed Thallium Odd-Mass Nuclei	A.M. Khalaf, M.M. Taha, and A.M. Ismail
017_ ICMEP-11	Design and Simulation of Terahertz Apodized Bragg Grating Using Coplanar Stripline Transmission Line a 1 μm-thin Membrane	Walid Gomaa and Thomas Darcie
018_ICMEP-11	Study the Optimal Operative Conditions of New Glow Discharge Ion Source for Polymeric Treatment	M.G. Hassan, B.A. Soliman, A.A. El-Saftawy, M.M. Abdelrahman, K.E. Abdelmageed and M.A. Hassouba

Journal of Physics: Conference Series **2304** (2022) 011001 doi:10.1088/1742-6596/2304/1/011001

Part II: Keynotes, Workshops and Seminars

Journal of Physics: Conference Series **2304** (2022) 011001 doi:10.1088/1742-6596/2304/1/011001

Military Technical College
Kobry El-Kobbah,
Cairo, Egypt.
icmep@mtc.edu.eg

11th International Conference on Mathematics and Engineering Physics
29-31 March 2022
https://icmep.conferences.ekb.eg/

Wednesday March 31, 2022 **Time 8:30 – 10:00**	**Session: WS I** **Room: A14**

Work Shops I

Chairman: Maj. Gen. Aymen Mokhtar **M.T.C.**
Brig. Dr. Mohamed Abdo **M.T.C.**

Left. Col. Walid Gomaa **M.T.C.**

WS 1_ ICMEP-11	Advances in Laser Remote Sensing of trace vapors in the atmosphere	Dr. Eng. Salah I. Hassab Elnaby NILES, Cairo University selnaby@niles.edu.eg
WS 2_ ICMEP-11	New trends: Quantum Computers	Dr. Eng. Salah I. Hassab Elnaby NILES, Cairo University selnaby@niles.edu.eg

Journal of Physics: Conference Series **2304** (2022) 011001 doi:10.1088/1742-6596/2304/1/011001

Military Technical College
Kobry El-Kobbah,
Cairo, Egypt.
icmep@mtc.edu.eg

11th International Conference on Mathematics and Engineering Physics
29-31 March 2022
https://icmep.conferences.ekb.eg/

Wednesday March 31, 2022	**Session: WS II**
Time: 10:15 – 11:45	**Room: A14**

Work Shops II

Chairman: Maj. Gen. Aymen Mokhtar M.T.C.
Brig. Dr. Mohamed El said M.T.C.
Brig.Dr. Mohamed A.E.Abdel-Rahman M.T.C.

WS 3_ ICMEP-11	Nuclear Security Culture	Major General Dr/Mohamed Saad Eldin Kamel (Head of Security Sector-Egyptian Atomic Energy Authority)
WS 4_ ICMEP-11	Peaceful Uses of Atomic Energy	Ass. Prof / Samah I. Radwan

Journal of Physics: Conference Series **2304** (2022) 011001 doi:10.1088/1742-6596/2304/1/011001

Military Technical College
Kobry El-Kobbah,
Cairo, Egypt.
icmep@mtc.edu.eg

11th International Conference on Mathematics and Engineering Physics
29-31 March 2022
https://icmep.conferences.ekb.eg/

Reviewers

Maj. Gen. Prof. Gamal Mabrouk Abdel-Hamid	Military technical collage
Maj. Gen. Prof. Hassan El-Hamouly	Military technical collage
Brig. Gen. Dr. Ahmed Salah Ismail	Military technical collage
Brig. Gen. Dr. Hany Nasry Zaky	Military technical collage
Col Dr. Ahmed Kamal Elshereif	Military technical collage
Col. Prof. Dr. Yasser H. Elsharkawy	Military technical collage
Brig. Assoc. Dr Ayman M. Mokhtar	Military technical collage
Maj. Gen. Prof. Kamal Abdelraham	British University
Prof. Dr. Mohamad Atta Abdu Elhaliem	National Institute of Laser Science
Prof. Dr. Mohamad Abdu Elhareth Mohamad	National Institute of Laser Science
Prof. Dr. Amer Mohamed Esmail Kany	Faculty of Science, Al-Azhar Univ
Prof. Dr. Ahmed Ashour	Egyptian Atomic Energy Authority (EAEA)
Prof. Dr. Labib Rashed	Ain Shams University
Prof. Dr. Mohamad Gharieb	Ain Shams University
Prof. Dr. Hatem El-Ghandoor	Faculty of Science - Cairo University
Prof. Dr. Salah I. Hassab Elnaby	Faculty of Science -Cairo University

Journal of Physics: Conference Series **2304** (2022) 011002 doi:10.1088/1742-6596/2304/1/011002

Peer Review Statement

All papers published in this volume have been reviewed through processes administered by the Editors. Reviews were conducted by expert referees to the professional and scientific standards expected of a proceedings journal published by IOP Publishing.

- **Type of peer review:** Double Anonymous
- **Conference submission management system:** https://icmep.conferences.ekb.eg/
- **Number of submissions received:** 24
- **Number of submissions sent for review:** 24
- **Number of submissions accepted:** 16
- **Acceptance Rate (Submissions Accepted / Submissions Received × 100):** 66.7
- **Average number of reviews per paper:** 2
- **Total number of reviewers involved:** 16
- **Contact person for queries:**
  Name: Prof. Dr. Gamal Mabrouk Abdelhamid,
  Email: gmabrouk@mtc.edu.eg
  Affiliation: Conference Rapporteur

Mathematical Model for Omnidirectional Sensor Network Using Clifford Algebra

Amr M. Mahfouz[1,4], **Hany N. Zaky**[1,5], **Ahmed S. Ismail**[1,2,6] **and Eiman T. Abou El Dahab**[3,7]

[1] Mathematics Department, Military Technical College, Cairo, Egypt.

[2] Head of Mathematics Department, Military Technical College, Cairo, Egypt.

[3] Faculty of Computer Science and Artificial Intelligence, Modern University for Technology and Information, Cairo, Egypt.

[4] amrmahfouz@mtc.edu.eg.

[5] hanynasry@mtc.edu.eg.

[6] A-s-Ismail@mtc.edu.eg.

[7] eiman@gmail.com.

Abstract. Wireless Sensor Networks are now very important and more usable in industrial, agricultural, commercial and military applications. Studies on wireless sensor networks have several aspects such as location, deployment, tracking. The most critical parameter of studying wireless sensor networks is sensor nodes deployment to obtain optimum coverage for the network used to minimize energy and cost. The goal of this work is to introduce a new coverage proposal to calculate the required number of sensors for a wireless sensor network in order to achieve optimum coverage for a particular task region. The other proposals that were introduced before are mostly based on coverage of a unit disk in two dimensional space or a ball in three dimensional space. These proposals didn't provide a coverage model for targets or objects in task region with hybrid types. So, this work introduces a new coverage model using Clifford algebra, which is coordinate-free and simple in calculations.

Keywords: Clifford Algebra, Wireless Sensor Network, Coverage.

1. Introduction

As new technologies which can transform the field of knowledge to new stages, intelligent wireless sensor networks (WSN) applications became a main research side, due to large variety of applications that need to use WSN. For example, hundreds of sensors are used in large modern buildings that are wired to a central air conditioning systems. These wiring cost and complex in environmental controls, that is why (WSN) are needed. Coverage problem in (WSN) is one of the most important issues that should be taken into consideration. Studies on WSN include distribution of the network, information gathering, intelligent information fusion, and so on [1]. Since the quality of service and information acquisition are affected by network distribution, study on distribution of the network is very critical. Therefore, the sensor network coverage problem is very important to ensure that the sensor nodes in the network cover the task region effectively to guarantee high quality of service (Qos) for the sensor network [2] [3]. Another sensor network problem is that the energy and the communication bandwidth are limited [4]. While a certain quality of space coverage is required, the coverage configuration is optimized to increase the lifetime of the network, reduce costs of nodes communication, and upgrade reliability of information transmission [5]. According to large redundancy in the applications of (WSN), a lot of coverage models are formulated according to the dimension of the environment (space) and the target needs to be detected.

There exist three types of coverage:

1- Point coverage, where a certain point is desired to be covered.
2- Area coverage, where some points in the same plane are covered.
3- Space coverage, in which the points desired to be covered are in three-dimensional space.

In case of studying the deployment sensors of the network mechanisms, it can be divided into "deterministic coverage", "random coverage", "movable coverage", and so on. Also, when studying methods of information acquisition, there is "geometric coverage" or "information coverage". The geometric coverage mainly related to the quality of coverage based on the geometric coverage model, that is a region centred by the node with distance equals to sensing radius [6]. The information coverage [7] mainly uses the collected information among sensor nodes to achieve coverage. Some task regions could be effectively covered with the "information coverage" method when geometrical coverage is unavailable.

In this work, geometric deterministic coverage in wireless sensor networks is studied using Clifford algebra which is a powerful mathematical tool that provides a direct and simple way to present the model of any phenomena geometrically. Recently Clifford algebra is applied in many research fields such as physics, robots and image processing. Because it presents vectors of different planes into a single mathematical system with a comprehensive geometric significance.

2. Related work.

In real world, it should be deal with three dimensional sensor nodes, whose coverage model and conformable collected information of the objects to be monitored are in three dimensional structure. Although, traditional two dimensional coverage algorithms are difficult to be used in real three dimensional Environment [8]. According to rapid research development and demand increase for practical applications in three dimensional physical world, more attention is paid for three dimensional (WSN), for example the acoustic wireless sensor network and wireless sensor network used to monitor atmospheric changes that are based mainly on three dimensional scene [9]. Recently, due to growing of underwater (WSN), researchers started to work on the sensor nodes' coverage model and its deployment scheme from the real physical world point of view. A distributed deployment method for underwater (WSN) is presented by Ammari and Das [8] to adjust nodes position under water in order to minimize the intersections between nodes. If the ratio between radius of communication for sensor nodes and sensing radius are equal or greater than a certain value, the coverage model represented using octahedron is preferable than the coverage model of cube [10]. The "Art Gallery Problem" is also one of the basic problems in sensor networks, since the number and the observers' positions must be calculated so that the art gallery room is covered. In order to guarantee that every point observed by one sensor node at least, the polygon should be divided into non-overlapping smaller triangles. In two dimensional space this problem can be solved optimally, but in three dimensional space it's not that easy. Generally, for large and complicated (WSN), there exist different types of the task regions to be covered, such as planar terrain, mountains, battlefield, jungle and other complex regions. so, in this paper coverage model based on Clifford algebra that provides a homogeneous, coordinate free coverage model for different dimensional spaces is presented.

3. Clifford Algebra (Geometric Algebra)

Clifford algebra, that is also known as geometric algebra, was first introduced by Clifford W K in 19th century [11]. Based on Gassmann algebra, it provides a homogeneous coordinate- free representation to obtain the relationships between objects in geometry in a space, it can be applied to the higher dimensional space easily. Now Clifford algebra is a crucial tool that can be applied in mathematics and physics applications [12].

3.1 Multiplication in Clifford algebra in R^3

Let us have the orthonormal basis $\{e_1, e_2, e_3\}$ of Euclidean vector space $\boldsymbol{R}^3$. two vectors a and b of $\boldsymbol{R}^3$ can be obtained as linear combinations of the basis elements:

$$a = a_1e_1 + a_2e_2 + a_3e_3 \quad (1)$$
$$b = b_1e_1 + b_2e_2 + b_3e_3 \quad (2)$$

By multiplying equation (1) and (2) we get:

$$ab = (a_1b_1 + a_2b_2 + a_3b_3) + (a_1b_2 - a_2b_1)e_1e_2 + (a_1b_3 + a_3b_1)e_1e_3 + (a_2b_3 + a_3b_2)e_2e_3 \quad (3)$$

In three dimension the product of two vectors results in scalar part and bivetor part. By multiplying the three basis vectors $e_1e_2e_3$ another element is introduced called trivector which presents the oriented volume In $\boldsymbol{R}^3$ [13].

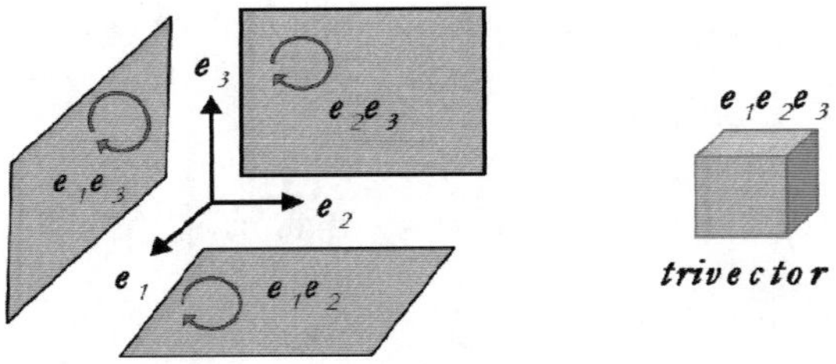

Figure.1 Oriented subspaces in three-dimensional Clifford algebra

According to that Clifford product gives a scalar part and trivector we can write the multiplication in Clifford as follows:

$$ab = a.b + a \wedge b \quad (4)$$

Where $a.b$ is called inner product and $a \wedge b$ is called outer product. The $\wedge$ (wedge) symbol is used to denote this operator that is also called wedge product.

3.2 Projections and rejections in Clifford algebra

We need to calculate the projections of a vector in space since it will be used in introducing the coverage model of wireless sensor network. The relative perpendicular and parallel components from the operations of projection are obtained, from equation (4), by reversing terms of vectors in geometric product:

$$ba = b.a + b \wedge a = a.b - a \wedge b \quad (5)$$

By adding and subtracting equations (4) and (5) we find that:

$$a.b = \frac{1}{2}(ab + ba) \quad (6)$$

$$a \wedge b = \frac{1}{2}(ab - ba) \quad (7)$$

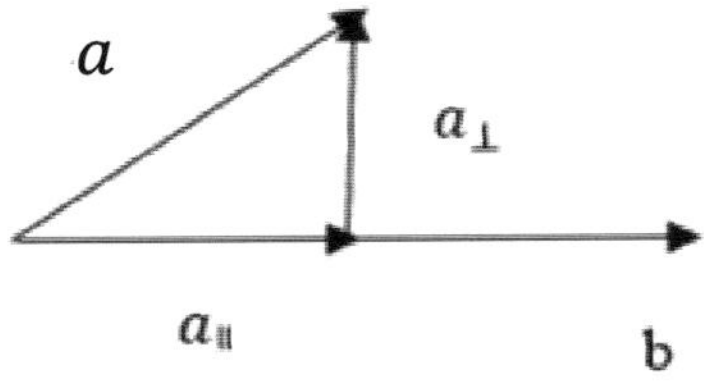

Figure 2. projection and rejection of $\boldsymbol{a}$ over $\boldsymbol{b}$

The parallel and perpendicular components are $a_\parallel = \frac{a.b}{b}$ and $a_\perp = \frac{a \wedge b}{b}$ respectively.

3.3 Rotations in three dimensional Clifford algebra

For a vector $\boldsymbol{v}$ having an angle with respect to plane p as shown in figure (3) that have a unit normal vector n, we can obtain the vector $\boldsymbol{v}'$ that is the reflection of $\boldsymbol{v}$ using Clifford algebra as follows [14]:

$$v = n^2v = n(n.v + n \wedge v) = (n.v)n + n(n \wedge v) \quad (8)$$

Since $\boldsymbol{v}_{\parallel}$ is the parallel component of $\boldsymbol{v}$ to plane P and $\boldsymbol{v}_{\perp}$ is the perpendicular component, where:

$$v = v_{\parallel} + v_{\perp} \tag{9}$$

By comparing equations (8) and (9) we find that the parallel and perpendicular components of the vector v are $v_{\parallel} = n(n \wedge v)$ and $v_{\perp} = (n.v)n$ respectively.

As a result the reflected vector v' for the plane vector P will be :

$$v' = v_{\parallel} - v_{\perp} = n(n \wedge v) - (n.v) = -n(n.v) - (n \wedge v) = -nvn \tag{10}$$

By doing the same thing to obtain the reflection about for the plane P_1 with another normal unit vector m will be:

$$v'' = -m(-nv'n)m = mnv'nm \tag{11}$$

From equation (11) the bivector mn is considered as a rotation operator $R = mn$, $\tilde{R} = nm$ where $R\tilde{R} = 1$.

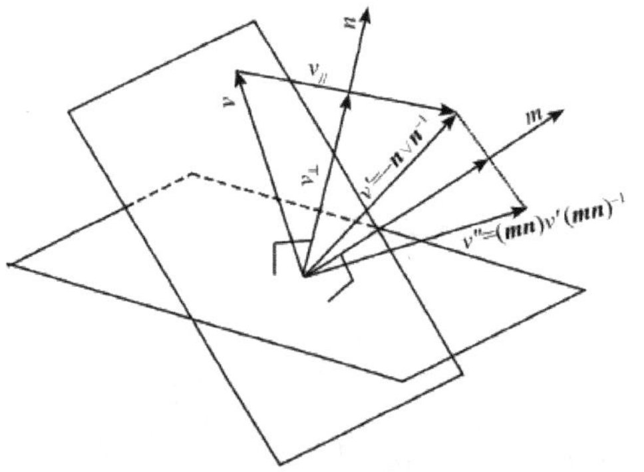

Figure 3. Rotation of a vector $\boldsymbol{v}$ in Clifford algebra

The rotating angle information is provided by the rotation operator R as it can be represented by $ae^{\theta I}$ where θ angle of rotation and a is the scale factor $a = \|R\|$ So $v' = Rv = ae^{\theta I}v$. Note that this property is important for the coverage analysis for wireless sensor networks.

4. Coverage analysis for (WSN) based on Clifford algebra

In this work, a new coverage proposal for (WSN) is introduced. This coverage model is based on Clifford rotation operator in which the geometric computations are efficient and more simple.

According to coverage definition in sensor networks, a certain point should be checked if it's detected or not.

For an omnidirectional sensor node S that have a sensing range ρ which is the length of the vector v_s, where v_s is the sensing vector of node S , the coverage range is $D = \{x: \|x - s\| \leq \rho\}$.

So if T is a specified region as show in figure 4 (a), it's discrimination model is:

$$C(T) = \begin{cases} 1, & \forall x \in T, \|R_x\| \leq 1 \\ 0, & \forall x \in T, \|R_x\| > 1 \end{cases} \tag{12}$$

Where $x = R_x v_s$

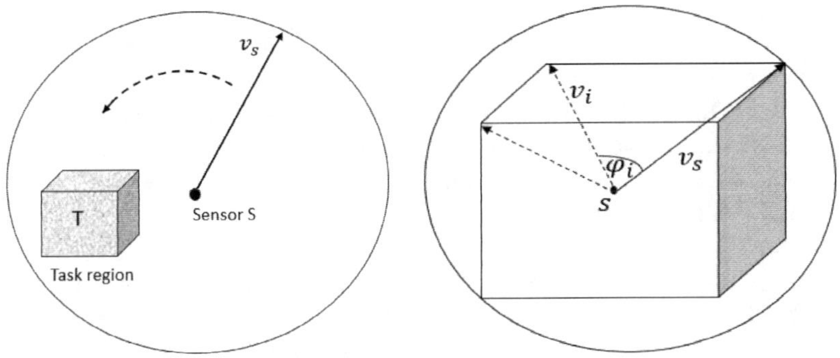

Figure 4. a) coverage model used for certain task region and b) cubic region using rotation operator

These two conditions of the coverage discrimination model represent that if the scale factor of the rotation operator is less than 1 and $\boldsymbol{v_s}$ is rotated to a certain point in the region of interest this point will be in the coverage range of sensor node S. On the other hand, if the scale factor of the rotation operator is more than 1 the specified task region will not be completely covered by the sensor node S.

The advantages of using the rotation operator of Clifford algebra in coverage analysis of sensor networks are:

1. It's coordinates-free, as it doesn't depend on a certain coordinates system.
2. It's also dimension-free that is it can be used to perform coverage analysis for any space dimensions.

According to the last advantages, an effective coverage analysis model is provided for wireless sensor networks that has hybrid types of targets.

For a sensor node S with the vector v_s oriented at this point as shown in figure 4(b) such that $\|v_s\| = \rho$, $v_s \neq v_i, \forall i = \{1,2,3, \dots \dots \dots .8\}$, and R_i is the rotation operator used to rotate a vector from v_s to v_i. And it's a necessary condition for the cube to be covered by the sensor node S that the scale factor of all the rotation operators is less or equal than 1, $\|R_i\| \leq 1$. Since $v_i = R_i v_s$, the rotation operator will be:

$$R_i = \frac{v_i}{v_s} = v_i(v_s(v_s.v_s)^{-1}) = \frac{v_i}{\|v_s\|}\frac{v_s}{\|v_s\|} \quad (13)$$

It's clear from equation (13) that if the cube is covered by S, so each vertex v_i will be covered by S, then $\|v_i\| \leq \|v_s\|$, so $\|R_i\| \leq 1, \forall i = \{1,2,3, \dots \dots \dots .8\}$. On the other hand if $\|R_i\| \leq 1 \ \forall i = \{1,2,3, \dots \dots \dots .8\}$ this will lead to $\|v_i\| \leq \|v_s\|$, and so all the vertices will be covered by S. Since this cube is a convex polygon, each point inside the cube will be in the coverage range of the node S.

Now, for a certain task region in the actual environment that it might be a forest for example, with the following dimensions: 100 m length, 100 m width, height of 20 m and a senor node of $r_s = 30$ m sensing range, we will subdivide this region into small cubes as shown in figure (5) with length of each cube equals to $\frac{2r_s}{\sqrt{3}}$. And a sensor node is put in the middle of each subdivided cube to achieve a full coverage of the actual region. According to calculations it takes around 5 sensors to achieve full coverage.

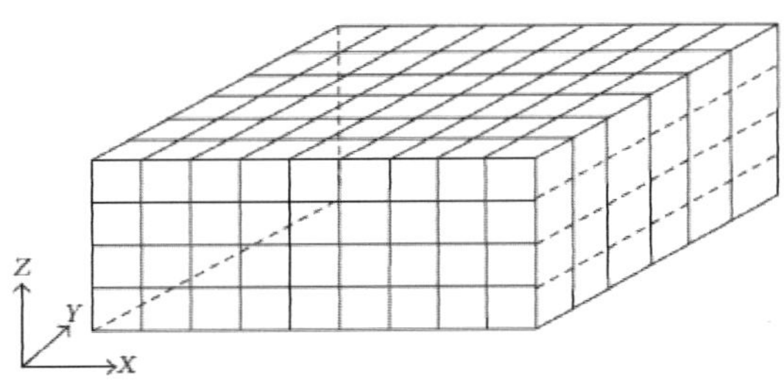

Figure 5. dissections of actual environment

5. Conclusion

In this paper, the main definitions and properties of Clifford algebra are discussed. Also, the main advantages for using the rotation operator of Clifford algebra were introduced. The wireless sensor network systems were studied from the deployment of sensors point of view, as we were interested in deterministic deployment. A new coverage proposal for achieving full coverage of wireless sensor networks is introduced using the rotation operator of Clifford algebra that provides coordinate-free implementation and it's simple in calculations. The effectivity of the model is verified by proving that a cube (task region) will be totally covered using sensor node. Finally, a practical example was introduced to explain the model and provide the number of sensors used to cover the given region.

References

[1] Jondhale, Satish R, Maheswar, 2022, "Fundamentals of Wireless Sensor Networks," in Received Signal Strength Based Target Localization and Tracking Using Wireless Sensor Networks, Springer, pp. 1-19.

[2] S. Megerian, F. Koushanfar, 2005, "Worst and best-case coverage in sensor networks," IEEE Trans Mobile Comput, vol. 4(1), pp. 84-92.

[3] S. Meguerdichian, M. Potkonjak, 2003, "Coverage problems in wireless ad-hoc sensor networks," IEEE Infocom, vol. 3, pp. 1380-1387.

[4] X. Li, O. Frieder, 2003, "Coverage in wireless ad hoc sensor networks," IEEE Trans Comput, vol. 52(6), pp. 753-763.

[5] Y. Ren, H. Zhang, 2006, "Three-dimensional optimal coverage routing protocol in wireless sensor networks," J Chinese, vol. 34(2), pp. 306-311.

[6] K. Alzoubi, X. Li, Y. Wang, 2003, "Geometric spanners for wireless ad hoc networks," IEEE Trans Parallel Distrib Syst, vol. 14(4), pp. 408-411.

[7] B. Wang, V. Srimivasan, 2006, "Sensor density for complete information coverage in wireless sensor network. In:," in Third European Workshop (EWSN). Lecture Notes in Computer Science, Berlin.

[8] M. Ammari, K. Das, 2009, "Critical density for coverage and connectivity in three-dimensional wireless sensor networks using continuum percolation," IEEE Transactions on Parallel, vol. 20(6), pp. 875-885.

[9] A. Newell, K. Akkaya, 2009, "Self-deployment of sensors for maximized coverage in underwater acoustic sensor networks," Computer Communications, vol. 32(7), pp. 1233-1244.

[10] S. Commuri, K. Watfa, 2006, "The 3-dimensional wireless sensor network coverage problem," in International Conference on Networking, Gainesville, Fla, USA.

[11] A. Macdonald, 2017, "A survey of geometric algebra and geometric calculus," Advances in Applied Clifford Algebras, vol. 27, no. 1, pp. 853-891.

[12] B. Li H, 2003, "Clifford algebra, geometric computing and reasoning," Adv Math, vol. 32(4), pp. 405-415.

[13] S. Mann, L. Dorst, 2002, "Geometric algebra: a computation framework for geometrical application, Part I," IEEE Comput Graph, vol. 22(3), pp. 24-31.

[14] H. Nasry, 2019, "Coordinate Transformation In Unmanned Systems Using Clifford Algebra," in Proceedings of the 5th International Conference on Mechatronics and Robotics Engineering, 2.

Study and Analysis of Corona-Virus Transfer Dynamics using Enhanced SEIR Epidemic Model with Vaccination Effect

Hazem M. Abd ElRaouf [1,4], Alhaytham M. Aref [2,5], Ahmed K. Elsherif [1,6] and Mohamed E. Khalifa[3,7]

[1]Department of Mathematics, Military Technical College, Cairo, Egypt.
[2]Egyptian Army
[3]Egyptian Chinese University, Cairo, Egypt.

[4]hazem.magdy108@mtc.edu.eg.
[5]Hmas_48@hotmail.com.
[6]ahmedelsherif80@gmail.com.
[7]khalifa@ecu.edu.eg.

Abstract. Corona Virus Syndrome (COVID-19) is a contagious disease and it is considered one of the deadliest viruses ever known to humanity. In this work, the transmission dynamics of the COVID- 19 has been studied using an enhanced SEIR epidemic compartmental model with a vaccination compartment. This model divides the whole population into five categories: susceptible (S), exposed (E), infectious (I), recovered (R), and vaccinated (V). Firstly, Positivity, Existence and Uniqueness of solution are illustrated. Secondly, a mathematical analysis is done to study the equilibrium points of the model and the basic reproduction number has been computed using Next Generation Matrix method. Thirdly, by using analytical computation the Stability of disease free equilibrium (DFE) point of the model is discussed and the Existence, and uniqueness of the endemic equilibrium are proved. Finally, Bifurcation coefficient is computed using the available data and Simulations are made to illustrate the effect of the transmission rate and vaccination rate on the value of the basic reproduction number.

Keywords: COVID-19, SEIR Model, Vaccination, Stability, Bifurcation, Basic Reproduction Number.

1. Introduction

The Corona-Virus infection (COVID-19) firstly recognized in December 2019 in a Chinese city called Wuhan city, and the virus has quickly spread over the world due to its high transmission rate. Specifically On 8 March 2020, the World Health Organization (WHO) broadcasted coronavirus as a global pandemic[1]. On 24 April 2020, more than 2.71 million conform cases have been reported in 110 countries. According to (WHO) till now more than "259 ,502, 031" cases were confirmed with "254,319,028" recovered and "5,183,003" deaths all over the world. Specifically, in Egypt we have about "355767" confirmed cases with "335462" recovered and "20305" deaths since the start of the pandemic[1]. The infectious diseases spread in a very complex manner that is influenced by a huge number of factors. With all of this intricacy, it's no surprise that simple mathematical models are frequently employed to describe the dynamics of infectious disease propagation. The adoption of more complicated models will necessarily include a huge number of difficult-to-estimate parameters, making model predictions weak and unpredictable. In this sense, compartmental models are widely employed to make infectious diseases mathematical modelling easier. The population is split into compartments in these models, with the premise that everyone in the same compartment has the same attributes [2]. One of the earliest models was proposed by Kermack and McKendrick [3] , who developed a SIR (susceptible-infectious-recovered) epidemic model in which the entire population was divided into three categories based on the individuals' disease condition. Following that, a variety of models, such as SEIR (susceptible-exposed-infectious-recovered) or SIS (susceptible-infectious-susceptible), were developed, analyzed, and applied to a range of infectious illnesses[4]. The SEIR mathematical model is a widely used compartmental epidemic model that involves dividing the total population into four fundamental compartments: susceptible (S), exposed to the disease but not yet infectious (E), infected (I), and removed (recovered or deceased) individuals (R). The SEIR model has been widely used to represent a variety of human infections, including Ebola [5] , H1N1[6] ,influenza [7] and MERS-CoV [8]. Vaccination is a very efficient way to prevent and reduce viral diseases [9]. Vaccination and effective control are critical elements in managing an epidemic as discussed in [10, 11].So In this work, to investigate the influence of vaccination on the transmission of COVID-19 infection, we employ an enhanced SEIR model with five compartments - susceptible ,exposed, infectious ,recovered, and vaccinated. The remainder of this work is structured as follows: In section 2, firstly we develop a new mathematical model and illustrate our mathematical equations by a Schematic diagram shown in figure (1) and also table (1) shows what every parameter in our equations refer to. In section 3, we present a mathematical analysis to demonstrate the positivity, existence and uniqueness of the solution of our model. then, in section 4 the disease free and the endemic equilibrium points of the model are illustrated and also the basic reproduction number which is the most essential parameter in studying the spread of infectious diseases has been computed using the technique of next generation matrix [12]. In section 5, we prove that the disease free equilibrium point is local asymptotic stable by computing all of the eigenvalues of the system's jacobian matrix, then we prove that when the value of the basic reproduction number surpasses unity, the endemic equilibrium point is accessible and locally asymptotically stable, whereas the disease-free equilibrium point is unstable. And we investigate the values of the bifurcation parameters in section 6, we used the available values for the parameters from local and global resources as shown later to calculate the values of both transmission rate (β) and vaccination rate (α) when the value of the basic reproduction number equals unity ,then we use MATLAB to simulate the variation of the basic reproduction number value with each of β and α values .Finally, Section 7 concludes our work, with an overview of the most important findings.

2. MATHEMATICAL MODEL

The flow of individuals through the five compartments of our model is represented via a set of Ordinary Differential Equations and Figure (1) shows the suggested model's disease transmission cycle.

$$\left.\begin{aligned}
\frac{dS}{dt} &= \mu N - \mu S - \beta SI - \alpha S \\
\frac{dE}{dt} &= \beta SI + \sigma\beta IV - \mu E - \varepsilon E \\
\frac{dI}{dt} &= \varepsilon E - \mu I - dI - rI \\
\frac{dR}{dt} &= rI + \gamma V - \mu R \\
\frac{dV}{dt} &= \alpha S - \sigma\beta IV - \mu V - \gamma V
\end{aligned}\right\} \xrightarrow{\text{Reduced to}} \left.\begin{aligned}
\frac{dS}{dt} &= \mu N - (\mu + \beta I + \alpha)\, S(t) \\
\frac{dE}{dt} &= \beta SI + \sigma\beta IV - (\mu + \varepsilon)E(t) \\
\frac{dI}{dt} &= \varepsilon E - (\mu + d + r)I(t) \\
\frac{dR}{dt} &= rI + \gamma V - \mu R \\
\frac{dV}{dt} &= \alpha S - (\sigma\beta I + \mu + \gamma)V(t)
\end{aligned}\right\} \quad (2.1)$$

S	Susceptible humans
E	Exposed humans
I	Infectious humans
R	Recovered humans
V	Vaccinated humans

Table 1. Explanation of Variables

μ	Natural "death /birth" rate
β	the transmission or contact rate
ε	The rate at which an exposed class becomes infected
d	death rate of I due to disease
r	Infectious class recovery rate
α	vaccination rate of population
σ	The vaccine inefficacy
γ	Rate of vaccinated people who turned to recovered class
N	Total population

Table 2. Explanation of parameters

Flow chart of the model:

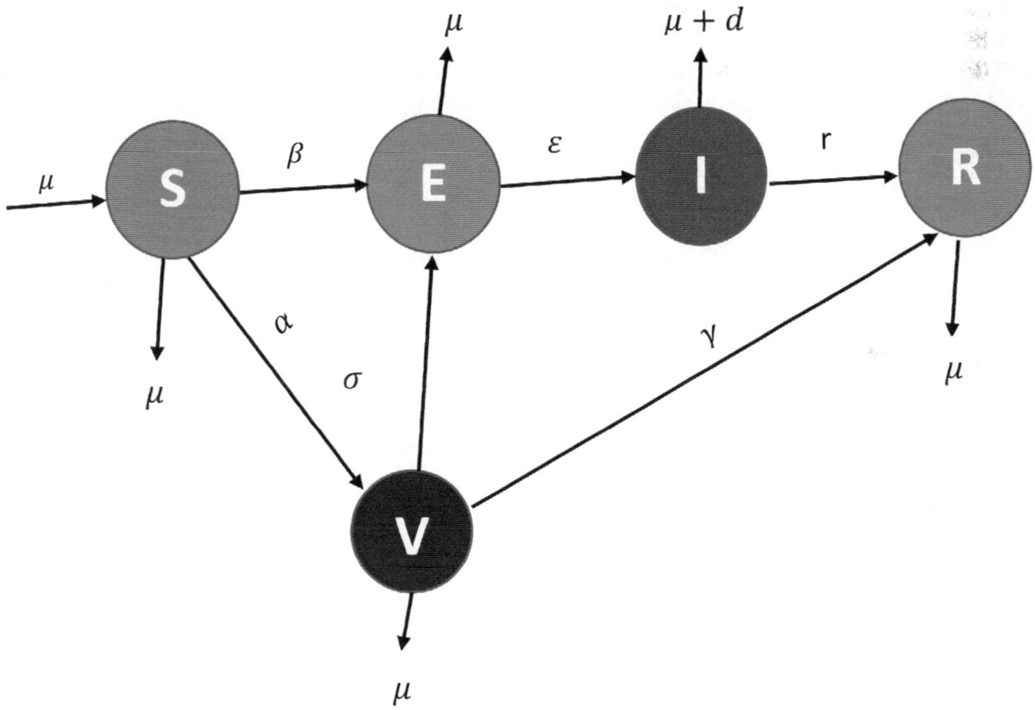

Figure (1): Schematic Diagram of Transmission of COVID-19

- The total population $N = S + E + I + R + V$.The vaccine inefficacy "σ" where $0 \leq \sigma \leq 1$ is a parameter which describes the vaccine efficiency. So $(1 - \sigma)$ reflects the vaccine's effectiveness. If $\sigma = 0$, the vaccination provides 100 percent resistance against the infection[13].

Susceptible (S): this class represents the individual who is at risk of becoming infected by COVID 19 virus. The number of susceptible individuals increases by the rate of natural birth (μ) and decreased by the rate of natural death (μ), transmission rate (β) and by vaccination rate of population (α).
Exposed (E): this class represents a member of a population who is infected and shows no clinical indications of the disease but he is cable of transferring it. This class increases by interaction with infected people at the rate (β) and with the vaccine inefficacy at the rate (σ). And it is decreased by the rate of natural death (μ) and the conversion rate of exposed individual to infectious class at the rate (ε).
Infectious (I): this class represents a member of a population who is infected with clinical symptoms of the virus and capable of transmitting it to others. The infectious persons' total number increases by the conversion rate from exposed individuals to infectious at the rate (ε). And decreased by each of recovery rate of infectious (r), disease induced death (d) and natural death (μ).
Recovered (R): this class represents an individual who recovered from the disease. Medical care of the infected population increases the number of people who have recovered at a rate (r) and also it is increased by vaccinated people who turned to recovered with the rate (γ), while natural death reduces it at the rate (μ).
Vaccinated(V): vaccinated individual is a member of population who took the vaccine ,this class of people is increased by vaccination rate of population (α) and decreased by each of (σ), (γ) and (μ).

3. Positivity of Solution, Existence and Uniqueness of solution

Theorem 1. For any t >0, all solutions of the set of equations (2.1) are nonnegative.
Proof : let $M = \{(S , E , I , V , R) \epsilon \mathbb{R}^5 : 0 \leq S,E,I,V,R\}$
To show that M is a set that is positively invariant, let us study the behaviors of the state variables in M.
a) At the boundary, $S = 0$, we get , $\frac{ds}{dt} = \mu N > 0$
so the solution is always positive and cannot cross the boundary $S = 0$.
b) At the boundary, $E = 0$, we have , $\frac{dE}{dt} = \beta SI + \sigma\beta IV$
if s > 0 , I>0 and V>0 then , $\frac{dE}{dt} \geq 0$
if I>0 ,V>0 and S=0 then , $\frac{dE}{dt} \geq 0$
if S>0, V>0 and I=0 then , $\frac{dE}{dt} = 0$
if S>0, I>0 and V=0 then , $\frac{dE}{dt} \geq 0$
In all cases $\frac{dE}{dt} \geq 0$, thus the boundary $E = 0$ can't be crossed by the solution.
c) At the boundary, I = 0, we have $\frac{dI}{dt} = \varepsilon E$
if E=0 then , $\frac{dI}{dt} = 0$
if E>0 then , $\frac{dI}{dt} > 0$
In all cases $\frac{dI}{dt} \geq 0$, so the boundary I = 0 can't be crossed by the solution.
d) At the boundary, R = 0, we have $\frac{dR}{dt} = rI + \gamma V$
if V=0 , I=0 then , $\frac{dR}{dt} = 0$
if V=0 , I>0 then , $\frac{dR}{dt} > 0$
if I=0 , V>0 then , $\frac{dR}{dt} > 0$

In all cases $\frac{dR}{dt} \geq 0$, so the boundary R = 0 can't be crossed by the solution. Therefore, the solution of the system of equations (2.1) is always in the set M and cannot cross the boundaries of any of our state variables. Assume that the following initial conditions are related to the system of equations (2.1):

$$S > 0, E \geq 0, V > 0, I > 0, R \geq 0 \qquad (3.1)$$

Theorem 2. consider the system of linear equations (2.1) which has nonnegative starting conditions (3.1). Then, for all $t \geq 0$ solutions of the system of equations (2.1) with its initial conditions (3.1) exist and they are unique.

Proof: Let $w(t) = (S, E, I, R, V) \epsilon \mathbb{R}^5$. Then the system of equations (2.1) could be written as $\frac{dw}{dt} = f(w)$. Let f_i represents the components of the vector field f for $i = 1,2,3,4,5$, we have,

$$\begin{aligned} f_1 &= \mu N - \mu S - \beta SI - \alpha S \\ f_2 &= \beta SI + \sigma\beta IV - \mu E - \varepsilon E \\ f_3 &= \varepsilon E - \mu I - dI - rI \\ f_4 &= rI + \gamma V - \mu R \\ f_5 &= \alpha S - \sigma IV - \mu V - \gamma V \end{aligned} \qquad (3.2)$$

The state variables' algebraic polynomials create the vector field f. So $f_i \epsilon \mathbb{R}^5$ are autonomous functions which are continuous and their partial derivatives $\frac{\partial f_i}{\partial S}, \frac{\partial f_i}{\partial E}, \frac{\partial f_i}{\partial I}, \frac{\partial f_i}{\partial R}, \frac{\partial f_i}{\partial V}$ exist and continuous. Hence, for every initial condition $w(0) \epsilon \mathbb{R}^5$, the system $\frac{dw}{dt} = f(w)$ has a unique solution which exists according to Existence and Uniqueness Theorem [14].

4. Equilibria of the system and Basic Reproduction Number

4.1. Equilibrium points of the system

There are two points of equilibrium for the system of equations (2.1); the disease free equilibrium point (P_0) and an endemic one (P_1). We get the points of equilibrium of the system by putting $\frac{ds}{dt} = 0$, $\frac{dE}{dt} = 0$, $\frac{dI}{dt} = 0$, $\frac{dR}{dt} = 0$, $\frac{dV}{dt} = 0$ i.e., by resolving the following system of equations:

$$\begin{aligned} 0 &= \mu N - (\mu + \beta I + \alpha) S \\ 0 &= \beta SI + \sigma\beta IV - (\mu + \varepsilon)E(t) \\ 0 &= \varepsilon E - (\mu + d + r)I(t) \\ 0 &= rI + \gamma V - \mu R \\ 0 &= \alpha S - (\sigma I + \mu + \gamma)V \end{aligned} \qquad (4.1)$$

A disease-free equilibrium is one which occurs when the disease does not spread. which is denoted by P_0 ; i.e., $E=I=0$. By solving (4.1), this point is one-of-a-kind and it has the following form:

$$P_0(S_0, E_0, I_0, R_0, V_0) = \left(\frac{\mu N}{\mu+\alpha}, 0, 0, \frac{\alpha\gamma N}{(\mu+\alpha)(\mu+\gamma)}, \frac{\alpha\mu N}{(\mu+\alpha)(\mu+\gamma)}\right) \qquad (4.2)$$

By solving (4.1) under the constraints that $S \neq 0$, $I \neq 0$, $E \neq 0$, $R \neq 0$, $V \neq 0$, another model equilibrium point called (endemic equilibrium) could be determined. This equilibrium one is unique for constant parameters of the model (2.1) and it takes the form $P_1 = (S_1, E_1, I_1, R_1, V_1)$ where

$$S_1 = \frac{\mu N}{\mu+\alpha+\beta I_1} \quad , \quad E_1 = \frac{\beta I_1(S_1+\sigma V_1)}{\mu+\varepsilon}$$
$$I_1 = \frac{\varepsilon I_1}{\mu+d+r} \quad , \quad R_1 = \frac{r I_1+\gamma V_1}{\mu} \quad , \quad V_1 = \frac{\alpha S_1}{\sigma I_1+\mu+\gamma} \tag{4.3}$$

4.2 Basic Reproduction Number

This parameter is among the most significant elements in studying epidemic models which is denoted by "R_0" and used to estimate a disease's transmission potential. IT is defined as the average number of new infections or additional cases caused by a solo infected person into a completely susceptible population[15]. R_0 can be calculated using a variety of methods. In this work , the next generation matrix approach will now be presented [12], where R_0 is the next generation matrix's spectral radius. If $R_0 > 1$, an epidemic is predicted to grow exponentially, and if $R_0 < 1$ the epidemic will stop.

So, NGM for this model is built as follows, initially, **F** and **V** are defined the transmission and transition function matrices

$$\mathfrak{F} = \begin{bmatrix} \beta SI \\ 0 \\ \beta\sigma IV \end{bmatrix} \qquad \mho = \begin{bmatrix} (\mu+\varepsilon)E \\ -\varepsilon E + (\mu+d+r)I \end{bmatrix} \tag{4.4}$$

and

$$F = \begin{bmatrix} 0 & \beta S_0 \\ 0 & 0 \\ 0 & \beta\sigma V_0 \end{bmatrix} \qquad V = \begin{bmatrix} \mu+\varepsilon & 0 \\ -\varepsilon & \mu+d+r \end{bmatrix} \tag{4.5}$$

Then,

$$FV^{-1} = \begin{bmatrix} \frac{\beta\varepsilon S_0}{(\mu+\varepsilon)(\mu+d+r)} & \frac{\beta S_0}{(\mu+d+r)} \\ \frac{\beta\varepsilon\sigma V_0}{(\mu+\varepsilon)(\mu+d+r)} & \frac{\beta\sigma V_0}{(\mu+d+r)} \end{bmatrix} \tag{4.6}$$

As a result, the basic reproductive number R_0 according to the disease-free equilibrium (4.2) has the formula:

$$R_0 = \frac{\beta(\varepsilon S_0+\varepsilon\sigma V_0+\mu\sigma V_0)}{(\mu+\varepsilon)(\mu+d+r)} \tag{4.7}$$

Finally, after substitution with (4.2) we get,

$$R_0 = \frac{\beta\mu N(\mu\varepsilon+\gamma\varepsilon+\varepsilon\sigma\alpha+\mu\sigma\alpha)}{(\mu+\varepsilon)(\mu+\gamma)(\mu+\alpha)(\mu+d+r)} \tag{4.8}$$

5. Equilibrium points Stability Analysis

5.1 Disease-free equilibrium point stability

Theorem : The model (2.1) has a disease free equilibrium(DFE) point which is locally asymptotically stable (LAS) if $R_0 < 1$ and if R_0 is greater than 1 it becomes unstable.

Proof: the system of equations (2.1) has the following Jacobian matrix about the DFE point

Journal of Physics: Conference Series **2304** (2022) 012002 doi:10.1088/1742-6596/2304/1/012002

$$J(P_0)=\begin{bmatrix} -a_1 & 0 & -\beta S_0 & 0 & 0 \\ 0 & -a_2 & \beta S_0+\sigma\beta V_0 & 0 & 0 \\ 0 & \varepsilon & -a_3 & 0 & 0 \\ 0 & 0 & r & -\mu & \gamma \\ \alpha & 0 & -\sigma\beta V_0 & 0 & -a_4 \end{bmatrix} \quad (5.1)$$

Where $a_1 = \mu+\alpha$, $a_2 = \mu+\varepsilon$, $a_3 = \mu+d+r$, $a_4 = \mu+\gamma$ (5.2)

We notice the sign of the real part of every eigenvalue for the Jacobian matrix J, if all of them have a negative real parts, then the DFE is locally asymptotically stable .and the DFE is unstable if any of them has a real part with positive sign.

The matrix (5.1) has five eigenvalues which are calculated by the well-known equation det $(\lambda I-J)=0$; i.e., the solutions of the characteristic equation below represent the eigenvalues of the system:

$$(a_1+\lambda)(a_4+\lambda)(\mu+\lambda)(\lambda^2+a_2\lambda+a_3\lambda+a_2a_3-\varepsilon\beta S_0-\varepsilon\beta\sigma V_0)=0 \quad (5.3)$$

So the eigenvalues are

$$\lambda_1=-a_1=-(\mu+\alpha) \quad , \quad \lambda_2=-a_4=-(\mu+\gamma) \quad , \quad \lambda_3=-\mu \quad (5.4)$$

$$\lambda_4=-\frac{1}{2}\left(a_2+a_3-\sqrt{(a_2+a_3)^2+4\varepsilon\beta\sigma V_0+4\varepsilon\beta S_0-4a_2a_3}\right) \quad (5.5)$$

$$\lambda_5=-\frac{1}{2}\left(a_2+a_3+\sqrt{(a_2+a_3)^2+4\varepsilon\beta\sigma V_0+4\varepsilon\beta S_0-4a_2a_3}\right) \quad (5.6)$$

Then by substituting with (4.7) in (5.5) and (5.6) we get the following formulas for λ_4 and λ_5

$$\lambda_4=-\frac{1}{2}\left(a_2+a_3-\sqrt{(a_2+a_3)^2+4a_2a_3(R_0-1)-4\mu\sigma V_0}\right) \quad (5.7)$$

$$\lambda_5=-\frac{1}{2}\left(a_2+a_3+\sqrt{(a_2+a_3)^2+4a_2a_3(R_0-1)-4\mu\sigma V_0}\right) \quad (5.8)$$

It is clear that when $R_0 < 1$, we have $\lambda_4 < 0$ and $\lambda_5 < 0$. Since all the Jacobian matrix's eigenvalues are negative, the DFE "P_0" is locally asymptotically stable. If $R_0 > 1$, there exists the eigenvalue λ_4 with positive real part. Hence, P_0 is unstable for $R_0 > 1$.[16]

5.2 The Endemic Equilibrium Point's Existence and Uniqueness

Here, the endemic equilibrium's existence and uniqueness are investigated when R0 > 1. Let the model (2.1) has an endemic equilibrium $P_1=(S_1, E_1, I_1, R_1, V_1)$ which is constructed by setting all the equations in model (2.1) to zero, getting the following linear system:

$$\begin{aligned} 0 &= \mu N-(\mu+\beta I_1+\alpha)S_1 \\ 0 &= \beta S_1 I_1+\sigma\beta I_1 V_1-(\mu+\varepsilon)E_1 \\ 0 &= \varepsilon E_1-(\mu+d+r)I_1 \\ 0 &= rI_1+\gamma V_1-\mu R_1 \\ 0 &= \alpha S_1-(\sigma I_1+\mu+\gamma)V_1 \end{aligned} \quad (5.9)$$

From the first, third and fifth equation of (5.9) we get S_1, E_1, V_1 in terms of I_1

$$S_1 = \frac{\mu N}{a_1+\beta I_1} \quad , \quad E_1 = \frac{a_3}{\varepsilon} I_1 \quad , \quad V_1 = \frac{\alpha\mu N}{(a_1+\beta I_1)(\sigma\beta I_1+a_4)} \tag{5.10}$$

Then by substituting with the values of S_1, E_1, and V_1 that shown above in the second line in (5.9), we can get a quadratic formula for the variable I_1:

$$b_2 {I_1}^2 + b_1 I_1 + b_0 = 0 \tag{5.11}$$

Where, $b_0 = a_1 a_2 a_3 a_4 (1 - R_0)$, $b_1 = a_2 a_3 \beta(\sigma a_1 + a_4) - \sigma\varepsilon\mu N\beta^2$, $b_2 = a_2 a_3 \sigma\beta^2$ (5.12)

And the values of a_1 , a_2 , a_3 , a_4 are as stated before in (5.2)

We notice that b_2 is positive number in all cases. and when $R_0 > 1$, b_0 is negative value. hence, when $R_0 > 1$ there is only one positive value for I_1 resulting a unique endemic equilibrium P_1.

6. Bifurcation analysis

when R_0<1,The DFE point P_0 attains a local asymptotic stability while it becomes unstable when $R0$>1. So, at P_0 the system attains a bifurcation at R_0=1, which is known as the backward bifurcation.by putting R_0=1 in (4.8), we have

$$\beta\mu N(\mu\varepsilon + \gamma\varepsilon + \varepsilon\sigma\alpha + \mu\sigma\alpha) = (\mu + \gamma)(\mu + \varepsilon)(\mu + d + r)(\mu + \alpha) \tag{6.1}$$

Parameter	Description	Value	Dimension	Reference
μ	Natural "death /birth" rate	$4.5 * 10^{-5}$	Per day	[17]
ε	Progression rate from exposed to infectious class	0.182	Per day	[18]
d	death rate of I due to disease	0.0582	Per day	[1]
r	Infectious class recovery rate	0.0686	Per day	[1]
α	Vaccination rate of population	$3.5 * 10^{-4}$	Per day	[19]
σ	The vaccine inefficacy	0.05	Dimensionless	[13]
γ	Rate of vaccinated people who turned to recovered class	0.075	Per day	[19]
N	Total population	$1.2 * 10^{8}$	Dimensionless	[17]

Table 3. parameters values

By substituting with parameters values of the model contained in above table in equation (6.1), we get $\beta = 1.1 * 10^{-8}$. It is a threshold value as if β< $1.1 * 10^{-8}$ then R_0<1 and the DFE point P_0 is asymptotic stable and if β> $1.1 * 10^{-8}$, the value of R_0 exceeds the unity and P_0 is unstable. Therefore, $\beta = 1.1 * 10^{-8}$ is the bifurcation value.

Likewise, we will repeat the same steps as above but this time with switching α instead of β and by setting the beta value that we get above we get $\alpha = 3.5 * 10^{-4}$ which also is considered as a bifurcation value, and the following figures shows the variation of the value of R_0 with each of transmission rate (β) and vaccination rate (α) .

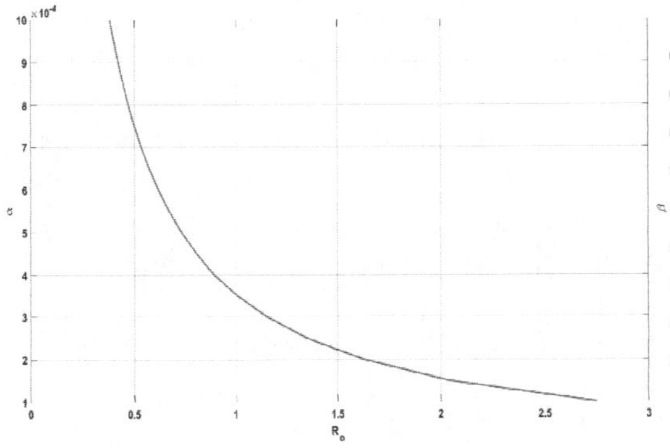

Figure (2): variation of R_0 with respect to α

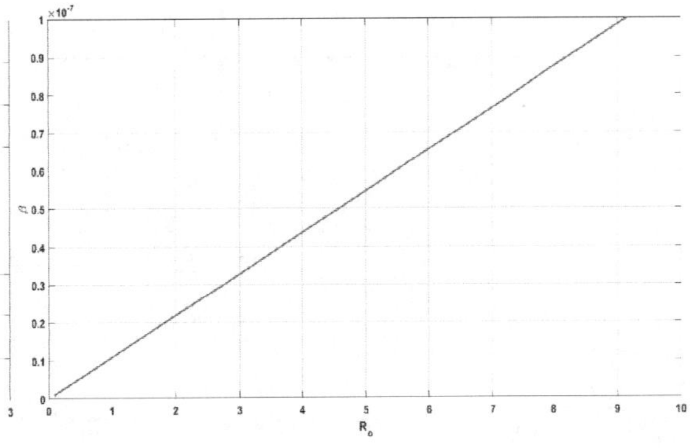

Figure (3): variation of R_0 with respect to β

7. Conclusion

Corona virus is currently serious issue in the world. This work presented an enhanced SEIR epidemic model containing five compartments, including vaccination, to analyze the disease's transmission dynamics. Initially, a mathematical model was built, then, the non-negativity, uniqueness and existence of the solution were proved, basic reproduction number of the model has been computed in terms of group of parameters, hence, we checked the disease-free equilibrium's stability and the endemic equilibrium's uniqueness. the system (2.1) was observed to have a bifurcation about the DFE point at the transmission rate $\beta = 1.1 * 10^{-8}$ and vaccination rate $\alpha = 3.5 * 10^{-4}$ which were considered to be the most effective model parameters, Increasing the transmission rate of disease over that limit increases the infectious population as a result of increasing the value of R_0 over 1,and Increasing the vaccination rate of population over that limit decreases the infectious population where R_0 value decreases below the unity. Thus, we can limit the spread of COVID-19 by raising the rate of vaccination of population and reducing the direct contact between the susceptible and infected individual.

8. References

[1] W. H. Organization, "World Health Organization coronavirus disease (COVID-19) dashboard," *World Health Organization,* 2020.

[2] O. Diekmann, H. Heesterbeek, and T. Britton, "Understanding Infectious Disease Dynamics. Princeton Series in Theoretical and Computational Biology," ed: Princeton University Press, 2013.

[3] W. O. Kermack and A. G. McKendrick, "A contribution to the mathematical theory of epidemics," *Proceedings of the royal society of london. Series A, Containing papers of a mathematical and physical character,* vol. 115, no. 772, pp. 700-721, 1927.

[4] M. Parsamanesh and M. Erfanian, "Global dynamics of an epidemic model with standard incidence rate and vaccination strategy," *Chaos, Solitons & Fractals,* vol. 117, pp. 192-199, 2018.

[5] P. Diaz, P. Constantine, K. Kalmbach, E. Jones, and S. Pankavich, "A modified SEIR model for the spread of Ebola in Western Africa and metrics for resource allocation," *Applied mathematics and computation,* vol. 324, pp. 141-155, 2018.

[6] M. M. Saito *et al.*, "Extension and verification of the SEIR model on the 2009 influenza A (H1N1) pandemic in Japan," *Mathematical biosciences,* vol. 246, no. 1, pp. 47-54, 2013.

[7] F. Etbaigha, A. R. Willms, and Z. Poljak, "An SEIR model of influenza A virus infection and reinfection within a farrow-to-finish swine farm," *PLOS one,* vol. 13, no. 9, p. e0202493, 2018.

[8] H.-J. Chang, "Estimation of basic reproduction number of the Middle East respiratory syndrome coronavirus (MERS-CoV) during the outbreak in South Korea, 2015," *Biomedical engineering online,* vol. 16, no. 1, pp. 1-11, 2017.

[9] R. S. Federman, "Understanding vaccines: a public imperative," *The Yale journal of biology and medicine,* vol. 87, no. 4, p. 417, 2014.

[10] G. B. Libotte, F. S. Lobato, G. M. Platt, and A. J. S. Neto, "Determination of an optimal control strategy for vaccine administration in COVID-19 pandemic treatment," *Computer methods and programs in biomedicine,* vol. 196, p. 105664, 2020.

[11] O. Sharomi and T. Malik, "Optimal control in epidemiology," *Annals of Operations Research,* vol. 251, no. 1-2, pp. 55-71, 2017.

[12] P. Van den Driessche, "Reproduction numbers of infectious disease models," *Infectious Disease Modelling,* vol. 2, no. 3, pp. 288-303, 2017.

[13] F. P. Polack *et al.*, "Safety and efficacy of the BNT162b2 mRNA Covid-19 vaccine," *New England Journal of Medicine,* 2020.

[14] S. Wiggins and M. Golubitsky, *Introduction to applied nonlinear dynamical systems and chaos*. Springer, 1990.

[15] O. Diekmann, J. A. P. Heesterbeek, and J. A. Metz, "On the definition and the computation of the basic reproduction ratio R 0 in models for infectious diseases in heterogeneous populations," *Journal of mathematical biology,* vol. 28, no. 4, pp. 365-382, 1990.

[16] M. Martcheva, *An introduction to mathematical epidemiology*. Springer, 2015.

[17] "Egyptian central agency for public Mobilization and statistics", ed.

[18] S. A. Lauer *et al.*, "The incubation period of coronavirus disease 2019 (COVID-19) from publicly reported confirmed cases: estimation and application," *Annals of internal medicine,* vol. 172, no. 9, pp. 577-582, 2020.

[19] "Egyptian Ministry of Health and Population."

Journal of Physics: Conference Series **2304** (2022) 012003 doi:10.1088/1742-6596/2304/1/012003

Applied method of characteristics on 2nd order linear P.D.E

Ahmed Saeed farg[1], A. M. Abd Elbary[2], Tarek A. Khalil[3]

[1]Department of Electrical Comminication and Electronic engineering , New Cairo Academy, Higher Institute of Engineering and Technology, New Cairo City, Egypt, ahmedgomea81@yahoo.com

[2]Department of Mathematics and Physical Science, New Cairo Academy, Higher Institute of Engineering and Technology, New Cairo City, Egypt, Dr_ah_abdelbary2005@yahoo.com

[3]Department of mathematics, New Cairo Academy, Higher Institute of Engineering and Technology, New Cairo City, Egypt, Tibrahimg@gmail.com

Abstract. PDEs are very important in dynamics, aerodynamics, elasticity, heat transfer, waves, electromagnetic theory, transmission lines, quantum mechanics, weather forecasting, prediction of crime places, disasters, how universe behave ……. Etc., second order linear PDEs can be classified according to the characteristic equation into 3 types hyperbolic, parabolic and elliptic; Hyperbolic equations have two distinct families of (real) characteristic curves, parabolic equations have a single family of characteristic curves, and the elliptic equations have none. All the three types of equations can be reduced to its first canonical form finding the general solution or the second canonical form similar to 3 basic PDE models; Hyperbolic equations reduce to a form coinciding with the wave equation in the leading terms, the parabolic equations reduce to a form modeled by the heat equation, and the Laplace's equation models the canonical form of elliptic equations. Thus, the wave, heat and Laplace's equations serve as basic canonical models for all second order linear PDEs.

1. Introduction

A PDE is an equation that contains one or more partial derivatives of an unknown function that depends on at least two variables. usually, one of these deals with time t and the remaining with space. PDEs are very important in dynamics, elasticity, heat transfer, electromagnetic theory, and quantum mechanics. The theory of partial differential equations of the second order is more complicated than the equations of the first order, and it is much more typical of the subject as a whole. Within the context, considerably better results can be achieved for equations of the second order in two independent variables than for equations in space of higher dimensions. Linear equations are the easiest to handle. In general, a second order linear partial differential equation is of the form

$$A(x,y)u_{xx} + B(x,y)u_{xy} + C(x,y)u_{yy} + D(x,y)u_x + E(x,y)u_y + F(x,y)u = G(x,y) \quad (1)$$

where A, B, C, D, E, F and G are in general functions of x and y but they may be constants. The subscripts are defined as partial derivatives where $u_x = \frac{\partial u}{\partial x}$

2. Canonical form

The classification of partial differential equations is suggested by the classification of the quadratic equation of conic sections in analytic geometry; $Ax^2 + Bxy + Cy^2 + Dx + Ey + f = 0$ represents hyperbola, parabola, or ellipse accordingly as B^2−4AC is positive, zero, or negative.
Classifications of PDE are:

(i) Hyperbolic if B^2−4AC > 0
(ii) Parabolic if B^2−4AC = 0
(iii) Elliptic if B^2−4AC < 0

The classification of second-order equations is based upon the possibility of reducing equation by coordinate transformation to canonical or standard form at a point. An equation is said to be hyperbolic, parabolic, or elliptic at a point (x_0, y_0) accordingly as; $B^2(x_0, y_0) - 4A(x_0, y_0)C(x_0, y_0)$ (2) is positive, zero, or negative. If this is true at all points, then the equation is said to be hyperbolic, parabolic, or elliptic. In the case of two independent variables, a transformation can always be found to reduce the given equation to canonical form in a given domain. However, in the case of several independent variables, it is not, in general, possible to find such a transformation
To transform equation (1) to a canonical form we make a change of independent variables. Let the new variables be; ε = ε (x, y), $\eta = \eta$ (x, y)
Assuming that ε and η are twice continuously differentiable and that the Jacobian; J= $\begin{vmatrix} \varepsilon_x & \varepsilon_y \\ \eta_x & \eta_y \end{vmatrix}$ is nonzero in the region under consideration, then x and y can be determined uniquely. Let x and y be twice continuously differentiable functions of ε and η Then we have,

$$u_x = \frac{\partial u}{\partial \varepsilon}\frac{\partial \varepsilon}{\partial x} + \frac{\partial u}{\partial \eta}\frac{\partial \eta}{\partial x} = u_\varepsilon \varepsilon_x + u_\eta \eta_x \quad , \quad u_y = \frac{\partial u}{\partial \varepsilon}\frac{\partial \varepsilon}{\partial y} + \frac{\partial u}{\partial \eta}\frac{\partial \eta}{\partial y} = u_\varepsilon \varepsilon_y + u_\eta \eta_y$$

$$u_{xx} = \frac{\partial u_x}{\partial \varepsilon}\frac{\partial \varepsilon}{\partial x} + \frac{\delta u_x}{\delta \eta}\frac{\delta \eta}{\delta x} = u_{\varepsilon\varepsilon}{\varepsilon_x}^2 + 2u_{\varepsilon\eta}\varepsilon_x\eta_x + u_{\eta\eta}{\eta_x}^2 + u_\varepsilon\varepsilon_{xx} + u_\eta\eta_{xx}$$

$$u_{yy} = \frac{\partial u_y}{\partial \varepsilon}\frac{\partial \varepsilon}{\partial y} + \frac{\partial u_y}{\partial \eta}\frac{\partial \eta}{\partial y} = u_{\varepsilon\varepsilon}{\varepsilon_y}^2 + 2u_{\varepsilon\eta}\varepsilon_y\eta_y + u_{\eta\eta}{\eta_y}^2 + u_\varepsilon\varepsilon_{yy} + u_\eta\eta_{yy}$$

$$u_{xy} = \frac{\partial u_x}{\partial \varepsilon}\frac{\partial \varepsilon}{\partial y} + \frac{\partial u_x}{\partial \eta}\frac{\partial \eta}{\partial y} = u_{\varepsilon\varepsilon}\varepsilon_x\varepsilon_y + u_{\eta\eta}\eta_x\eta_y + u_\varepsilon\varepsilon_{xy} + u_\eta\eta_{xy} + u_{\varepsilon\eta}(\varepsilon_x\eta_y + \varepsilon_y\eta_x)$$

substituting in (1)

$$A^*(x,y)u_{xx} + B^*(x,y)u_{xy} + C^*(x,y)u_{yy} + D^*(x,y)u_x + E^*(x,y)u_y + F^*(x,y)u = G^*(x,y) \quad (3)$$

Where; $A^* = A{\varepsilon_x}^2 + B\varepsilon_x\varepsilon_y + c{\varepsilon_y}^2$, $B^* = 2A\varepsilon_x\eta_x + B(\varepsilon_x\eta_y + \varepsilon_y\eta_x) + 2C\varepsilon_y\eta_y$
$C^* = A{\eta_x}^2 + B\eta_x\eta_y + C{\eta_y}^2$, $D^* = A\varepsilon_{xx} + B\varepsilon_{xy} + C\varepsilon_{yy} + D\varepsilon_x + E\varepsilon_y$
$E^* = A\eta_{xx} + B\eta_{xy} + C\eta_{yy} + D\eta_x + E\eta_y$, $F^* = F$, G^* =G
The resulting equation (3) is in the same form as the original equation (1) under the general transformation. The nature of the equation remains constant if the Jacobian does not vanish.
${B^*}^2 - 4A^*C^* = J^2(B^2 - 4AC)$ and $J^2 \neq 0$, We shall assume that the equation under consideration is of the single type in a given domain. The classification of equation (1) depends on the coefficients A (x, y), B (x, y), and C (x, y) at a given point (x, y) so equation (1) rewritten as;
$A(x,y)u_{xx} + B(x,y)u_{xy} + C(x,y)u_{yy} = H(x,y,u,u_x,u_y)$ (4) where; A, B, C≠0
And equation (3) rewritten as;
$A^*(x,y)u_{\varepsilon\varepsilon} + B^*(x,y)u_{\varepsilon\eta} + C^*(x,y)u_{\eta\eta} = H(\varepsilon,\eta,u,u_\varepsilon,u_\eta)$ Where A^*, C^* =0
$A{\varepsilon_x}^2 + B\varepsilon_x\varepsilon_y + c{\varepsilon_y}^2 = 0$, $A{\eta_x}^2 + B\eta_x\eta_y + C{\eta_y}^2 = 0$
Since the 2 equations from the same type, we can rewrite them;
$A{\varepsilon_x}^2 + B\varepsilon_x\varepsilon_y + c{\varepsilon_y}^2 = 0$ where $\boldsymbol{\varepsilon}$ stands for the 2 functions ε, η
Dividing by ${\varepsilon_y}^2$ $A(\frac{\varepsilon_x}{\varepsilon_y})^2 + B\frac{\varepsilon_x}{\varepsilon_y} + C = 0$ since in partial differentiation; $\frac{dy}{dx} = -\frac{\varepsilon_x}{\varepsilon_y}$

$A\left(\frac{dy}{dx}\right)^2 - B\frac{dy}{dx} + C = 0$; therfore two roots are $\frac{dy}{dx} = \frac{B \pm \sqrt{B^2 - 4AC}}{2A}$

These equations, which are known as the characteristic equations, are ordinary differential equations for families of curves in the by-plane along which $\boldsymbol{\varepsilon}$ = constant and η = constant. The integrals of equations

are called the characteristic curves. Since the equations are first order ordinary differential equations, the solutions may be written as; $\Phi_1(x, y) = c_1$ $\Phi_2(x, y) = c_2$ with c_1 and c_2 as constants.
Hence the transformations $\varepsilon = \Phi_1(x, y)$, $\eta = \Phi_2(x, y)$ will transform equation (4) to a canonical form.
We show that the characteristic of any hyperbolic PDE can be transformed as;
$*B^2 - 4AC > 0$ so, we have 2 real different characteristic integration yields reduced into first canonical form $u_{\varepsilon\eta} = H(\varepsilon, \eta, u, u_\varepsilon, u_\eta)$, $B^* \neq 0$
let we have new independent variable α, β where $\alpha = \varepsilon + \eta$, $\beta = \varepsilon - \eta$ to eliminate $u_{\varepsilon\eta}$ by setting $B^*(x, y)u_{\varepsilon\eta}=0$ where $u_{\varepsilon\eta} \neq 0$ so $A^* = -C^*$ $B^* = 2A\varepsilon_x\eta_x + B(\varepsilon_x\eta_y + \varepsilon_y\eta_x) + 2C\varepsilon_y\eta_y = 0$
which is transformed into second canonical form $u_{\alpha\alpha} - u_{\beta\beta} = H(\alpha, \beta, u, u_\alpha, u_\beta)$ similar to wave equation to be modeled
$*B^2 - 4AC = 0$ so, we have 2 real repeated characteristic integration yields reduced into canonical form $u_{\varepsilon\varepsilon} = H(\varepsilon, \eta, u, u_\varepsilon, u_\eta)$
$*B^2 - 4AC < 0$ so, we have no real characteristic but it has complex solution which is analytic along some neighborhood domain can be reduced into first canonical form
$u_{\varepsilon\eta} = H(\varepsilon, \eta, u, u_\varepsilon, u_\eta)$ where $\varepsilon = \alpha + i\beta$, $\eta = \alpha - i\beta$ are two conjugate functions where;
$\alpha = \frac{1}{2}(\varepsilon + \eta)$, $\beta = \frac{1}{2i}(\varepsilon - \eta)$ which can be used to eliminate $u_{\varepsilon\eta}$ by setting $B^*(x, y)u_{\varepsilon\eta} = 0$ where $u_{\varepsilon\eta} \neq 0$, so $B^* = 0$ $A^* = C^*$ $B^* = 2A\varepsilon_x\eta_x + B(\varepsilon_x\eta_y + \varepsilon_y\eta_x) + 2C\varepsilon_y\eta_y = 0$
Which is transformed into second canonical form $u_{\alpha\alpha} + u_{\beta\beta} = H(\alpha, \beta, u, u_\alpha, u_\beta)$ similar to la place equation to be modeled

3. Hyperbolic equations

3.1. Homogeneous wave equation

$u_{tt} - c^2u_{xx} = 0$, $B^2 - 4AC = 4c^2 > 0$, $A\lambda^2 - B\lambda + c = 0$

$\frac{dt}{dx} = \frac{\pm 1}{c}$ *separation of variables and integrate we get* $\varepsilon = x + ct$, $\eta = x - ct$

$u_{xx} = u_{\varepsilon\varepsilon}\varepsilon_x{}^2 + 2u_{\varepsilon\eta}\varepsilon_x\eta_x + u_{\eta\eta}\eta_x{}^2 + u_\varepsilon\varepsilon_{xx} + u_\eta\eta_{xx}$

$u_{tt} = u_{\varepsilon\varepsilon}\varepsilon_t{}^2 + 2u_{\varepsilon\eta}\varepsilon_t\eta_t + u_{\eta\eta}\eta_t{}^2 + u_\varepsilon\varepsilon_{tt} + u_\eta\eta_{tt}$

Then substitute in original P.D.E $-4c^2u_{\varepsilon\eta} = 0$, $c \neq 0$, $u_{\varepsilon\eta} = 0$ *the integrate w.r.t* η

$u_\varepsilon = f(\varepsilon)$ *then integrate w.r.t* ε $u = f(\varepsilon) + g(\eta) = f(x + ct) + g(x - ct)$

Using initial and boundary conditions $f(x + ct)$, $g(x - ct)$ can be determined

$u_{(x,0)} = p(x)$, $u_{t(x,0)} = v(x)$, $u_{(0,t)} = u_{(l,t)} = 0$, $t > 0$

$f(x) + g(x) = p(x)$, $f^\backslash(x) - g^\backslash(x) = \frac{1}{c}v(x)$ *by integration w.r.t x*

$f(x) - g(x) = \frac{1}{c}\int_{x_0}^{x} v(x)\,dx + f(x_0) - g(x_0)$ *solving for* $f(x)$, $g(x)$

$f(x + ct) = \frac{1}{2}p(x + ct) + \frac{1}{2c}\int_{x_0}^{x+ct} v(x)\,dx + \frac{1}{2}(f(x_0) - g(x_0))$

$g(x - ct) = \frac{1}{2}p(x - ct) - \frac{1}{2c}\int_{x_0}^{x-ct} v(x)\,dx - \frac{1}{2}(f(x_0) - g(x_0))$

$u_{(x,t)} = \frac{1}{2}(p(x + ct) + p(x - ct)) + \frac{1}{2c}\int_{x-ct}^{x+ct} v(x)\,dx$ *(De alembart)*

Where $t > 0$, $0 \leq x + ct \leq l$, $0 \leq x - ct \leq l$, $t \leq \frac{l-x}{c}$, $t \leq \frac{x}{c}$

$u_{(0,t)} = u_{(l,t)} = 0$, $t > 0$, $f(ct) = -g(-ct)$, $-ct \leq 0$

$f(l + ct) = -g(2l - (l + ct))$, $l + ct \geq l$ f, g *odd periodic*

$-g(2l - (l + ct)) = -\frac{1}{2}p(2l - (l + ct)) + \frac{1}{2c}\int_0^{2l-(l+ct)} v(x)\,dx + \frac{1}{2}(f(0) - g(0))$

$l \leq x + ct \leq 2l$ *extended to the right as shown in fig 1.a*

$l + ct = -(l - ct)$ $f(-(l - ct)) = \frac{1}{2}p(-(l - ct)) + \frac{1}{2c}\int_0^{-(l-ct)} v(x)\,dx + \frac{1}{2}(f(0) - g(0))$

$-2l \leq x - ct \leq -l$ *extended to the left as shown in fig 1.b*

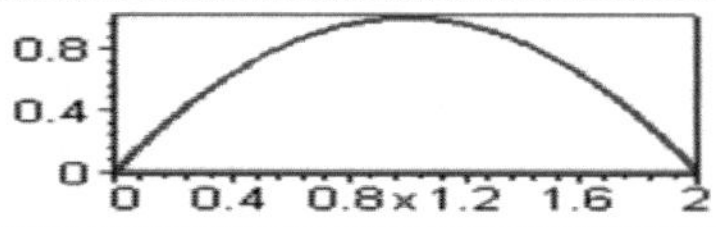

Figure 1. a string of length 2 length unit

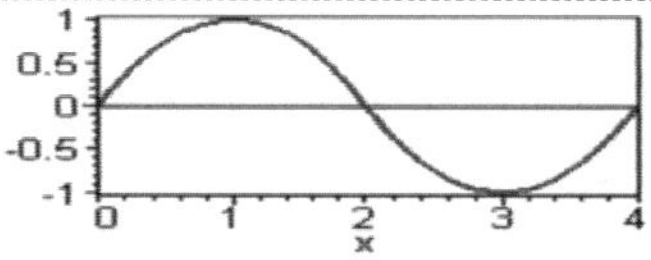

Figure 1.a The string extended to the right

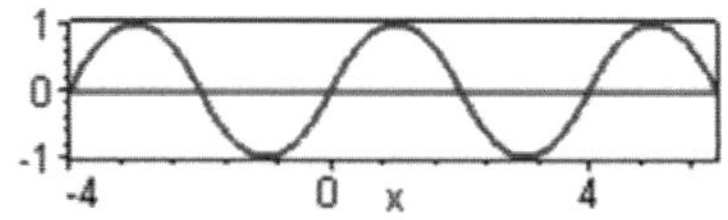

Figure 1.b The string extended to the left

3.1.1. physical application

1-Motion of stretched string in musical instruments such as guitar, piano described by $u_{tt} - c^2 u_{xx} = 0$ $where \;\; c^2 = \frac{T}{\rho}$ T horizantel component of tension force, ρ mass per unit length Suppose a such string placed on x-axis

I. Damping forces are neglected such as air resistance

II. Weight of string is also neglected

III. Tension force is tangential to string curve

Initial position function $u_{(x,0)} = p(x) = \sin\left(\frac{n x \pi}{l}\right)$ $l \; is \; length \; of \; string$

Initial velocity function $u_{t_{(x,0)}} = v(x) = 0$ (initially at rest), Boundaries $u_{(0,t)} = u_{(l,t)} = 0 \;, \; t > 0$

applying de Alembert formula $u_{(x,t)} = \frac{1}{2}\big(p(x+ct) + p(x-ct)\big) + \frac{1}{2c}\int_{x-ct}^{x+ct} v(x)\, dx$

$u_{(x,t)} = \frac{1}{2}\left(\sin\left(\frac{n\pi}{l}(x+ct)\right) + \sin\left(\frac{n\pi}{l}(x-ct)\right)\right)$ $\quad let \;\; n = 1 \,, \;\; l = c = 2$

$u_{(x,t)} = \frac{1}{2}\left(\sin\left(\frac{\pi}{2}(x+2t)\right) + \sin\left(\frac{\pi}{2}(x-2t)\right)\right) = \sin\left(\frac{\pi x}{2}\right)\cos(\pi t) \;\; as \; shown \; in \; fig. 2$

By increasing n $n = 2 \,, \;\; l = c = 2 \;\; , \quad u_{(x,t)} = sin(\pi x)\, co\, s(2\pi t) \;\; as \; shown \; in \; fig. 3$

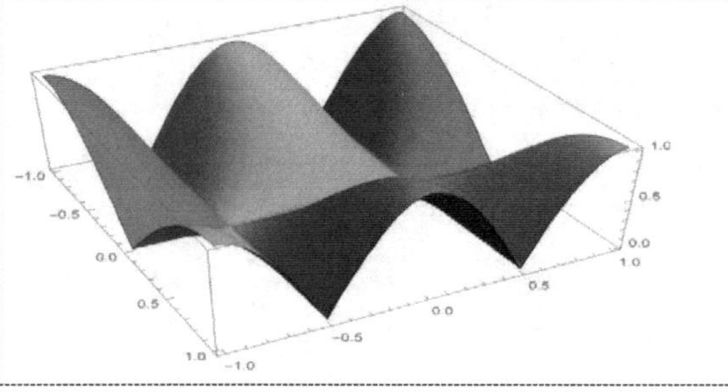

Figure 2.a. *complex plot for x, u and t.*

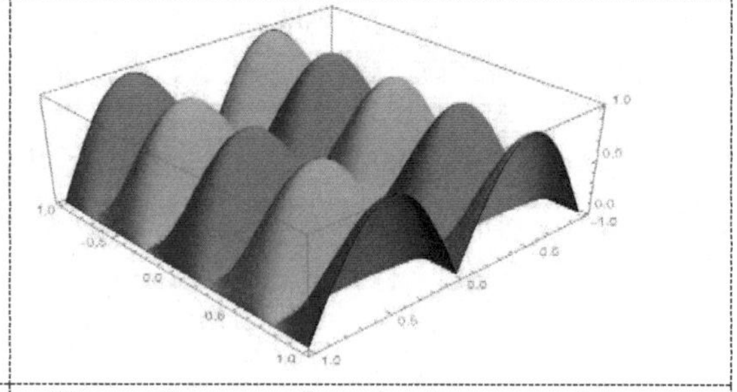

Figure 3.a. *complex plot for x, u and t*

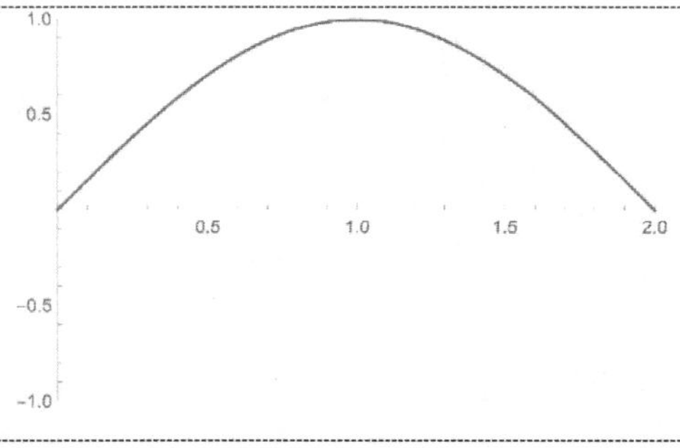

Figure 2.b. *2D Real plot for u and x.*

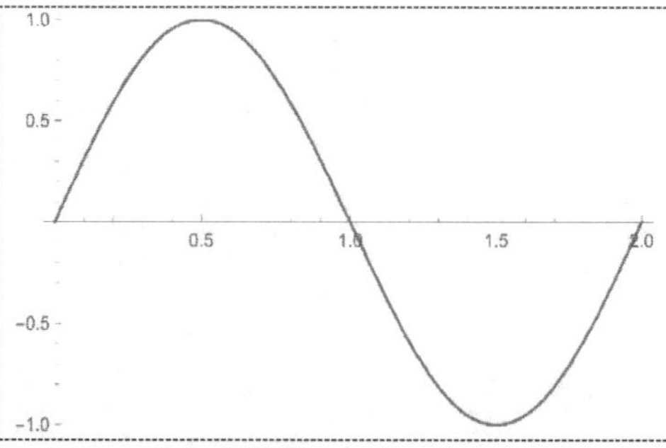

Figure 3.b. *2D Real plot for u and x..*

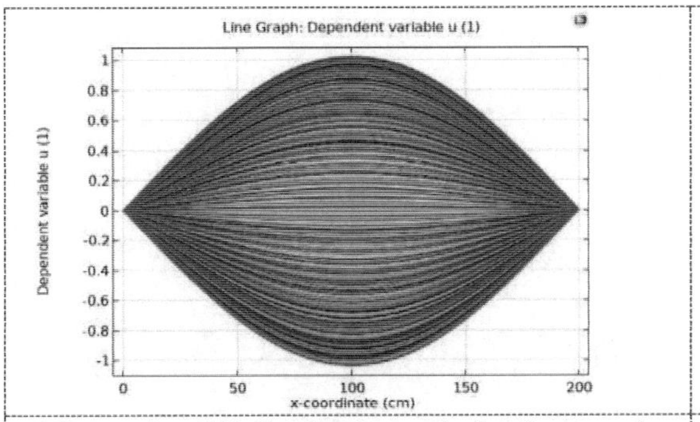

Figure 2.c. *2D Real animated plot for u and x.*

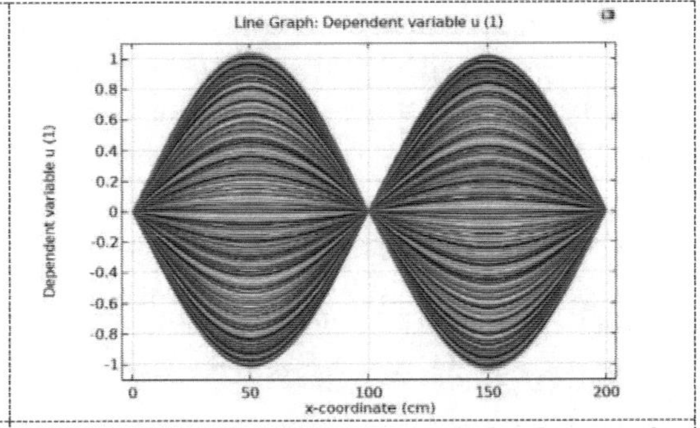

Figure 3.c. *2D Real animated plot for u and x*

2- Longitudinal waves travelling along thin Rod with Youngs Y modulus and mass density ρ where the constant $c^2 = \frac{Y}{\rho}$ is phase velocity where c is specific for each material.

Table: Calculated and measured longitudinal wave speeds in thin rods made up of common metals. Sources Haynes and Lide 2011.

Metal	Y (M m^{-2})	P (kg m^{-3})	$(Y/p)^{-1/2}$ (m s^{-1})	v (m s^{-1})
Aluminum	7.0×10^{10}	2.7×10^{3}	5100	5000
Copper	1.2×10^{11}	8.9×10^{3}	3600	3800
Lead	1.6×10^{10}	1.1×10^{4}	1100	1100
Nickel	2.0×10^{11}	8.9×10^{3}	4700	4900
Silver	8.3×10^{10}	1.1×10^{4}	2800	2700
Tin	5.0×10^{10}	7.4×10^{3}	2600	2700
Zinc	1.1×10^{11}	7.1×10^{3}	3900	3900

3-high frequency AC lossless cable (optical fiber, submarine cable, and transmission lines) where;
the cable is made such that resistance R and leakage of conductance G is also neglected
the general telegraph equation $i_{xx} = LCi_{tt} + (RC + GL)i_t + RGi$, $R = G = 0$
L inductance, C capacitance, R resistance, $i_{tt} - \frac{1}{LC} i_{xx} = 0$ high freq. AC similar to wave equation

$$i(x,t) = f\left(x + \frac{t}{\sqrt{LC}}\right) + g\left(x - \frac{t}{\sqrt{LC}}\right) \ , \ V(x,t) = f\left(x + \frac{t}{\sqrt{LC}}\right) + g\left(x - \frac{t}{\sqrt{LC}}\right)$$

3.2. Non-homogeneous wave equation

$$u_{tt} - c^2 u_{xx} = h^*(x,t) \ , \ u_{(x,0)} = p(x) \ , \ u_{t(x,0)} = v^*(x) \quad let\ y = ct$$

$$u_{xx} - u_{yy} = h(x,y) \ , \ u_{(x,0)} = p(x) \ , \ u_{y(x,0)} = v(x) = \frac{v^*(x)}{c} \ , \ h(x,y) = -\frac{h^*(x,t)}{c^2}$$

Let 2 characteristic curves ε, η where $\varepsilon = x - y$, $\eta = x + y$ as ε, η are constants so the 2 characteristic lines intersect at point p_0 and x axis at p_1, p_2 where the area is regioned by 3 lines B_0, B_1, B_2 as shown in fig. (4)

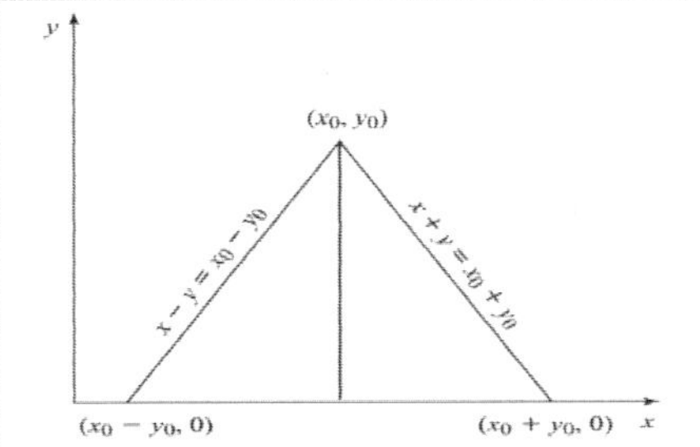

Figure 4.a. triangular region

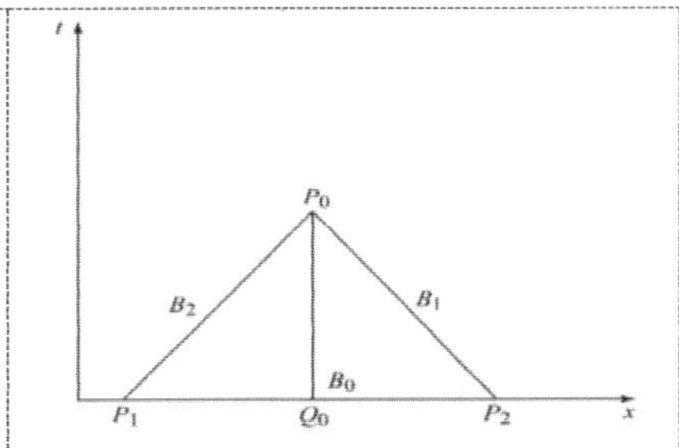

Figure 4.b. triangular region

As the area represents a double integral along the interior surface of triangle to get u where;

$\iint u_{xx} - u_{yy}\, dA = \iint h(x,y)\, dA$ applying greens theorem

$\iint (Q_x - P_y)\, dxdy = \oint Pdx + Qdy$, *the closed integral along B* , *as* $dA = dxdy$

$\iint u_{xx} - u_{yy}\, dA = \oint u_y dx + u_x dy$ *along* B_0, B_1, B_2

$\oint u_y dx + u_x dy$, *along* B_0 , *where* $y = 0$, $dy = 0$, $\oint u_y dx + u_x dy = \int_{x_o - y_o}^{x_o + y_o} u_y\, dx$

$\oint u_y dx + u_x dy$, *along* $B_1 : x = x_o + (y_o - y)$, $dx = -dy$

$\oint u_y dx + u_x dy = \oint -u_y dy - u_x dx = \int_0^{y_o} -u_y dy + \int_{x_o+y_o}^{x_o} -u_x dx = u_{(x_o+y_o,0)} - u_{(x_o,y_o)}$

$\oint u_y dx + u_x dy$, *along* $B_2 : x = x_o - (y_o - y)$, $dx = dy$

$\oint u_y dx + u_x dy = \oint u_y dy + u_x dx = \int_{y_o}^{0} u_y dy + \int_{x_o}^{x_o - y_o} u_x dx = u_{(x_o-y_o,0)} - u_{(x_o,y_o)}$

$\iint u_{xx} - u_{yy}\, dA = u_{(x_o-y_o,0)} + u_{(x_o+y_o,0)} - 2u_{(x_o,y_o)} + \int_{x_o-y_o}^{x_o+y_o} u_y\, dx$

$\iint h(x,y)\, dA = u_{(x_o-y_o,0)} + u_{(x_o+y_o,0)} - 2u_{(x_o,y_o)} + \int_{x_o-y_o}^{x_o+y_o} u_y\, dx$

$u_{(x_o,y_o)} = \frac{1}{2}\left(u_{(x_o-y_o,0)} + u_{(x_o+y_o,0)}\right) + \frac{1}{2}\int_{x_o-y_o}^{x_o+y_o} u_y\, dx - \frac{1}{2}\iint h(x,y)\, dxdy$

$u_{(x,0)} = p(x),\ u_{y_{(x,0)}} = v(x)$

$u_{(x,y)} = \frac{1}{2}(P(x-y) + P(x+y)) + \frac{1}{2}\int_{x_o-y_o}^{x_o+y_o} v(x)\, dx - \frac{1}{2}\int_0^{y_o}\int_{x_o-(y_o-y)}^{x_o+(y_o-y)} h(x,y) dxdy$

3.2.1. Physical application

1-Motion of stretched string in musical instruments such as guitar, piano …. etc. described by $u_{tt} - c^2 u_{xx} = h^*(x,t)$ *where* $c^2 = \frac{T}{\rho}$, T horizantel component of tension force, ρ mass per unit length, $h^*(x,y)$ are damping forces due to weight of string or air resistance Suppose a such string placed on x-axis, Initial position function $u_{(x,0)} = p(x) = 0$, *Damping of* $h^*(x,t) = xe^{-t}$

Initial velocity function $u_{t_{(x,0)}} = v(x) = (1 - \cos(x))e^{-x}$, $c = 1$

$u_{tt} - u_{xx} = xe^{-t}$, *applying the previous formula* $u_{xx} - u_{yy} = -xe^{-y}$

$u_{(x,y)} = \frac{1}{2}(P(x-y) + P(x+y)) + \frac{1}{2}\int_{x_o-y_o}^{x_o+y_o} v(x)\, dx - \frac{1}{2}\int_0^{y_o}\int_{x_o-(y_o-y)}^{x_o+(y_o-y)} h(x,y) dxdy$

$u_{(x,y)} = \frac{1}{2}\int_{x_o-y_o}^{x_o+y_o} (1-\cos(x))e^{-x}\, dx + \frac{1}{2}\int_0^{y_o}\int_{x_o-(y_o-y)}^{x_o+(y_o-y)} xe^{-y}\, dx\, dy$ as shown in fig. (5)

$$u = \frac{e^{-x-t}}{4}(2 - \mathrm{Cos}[x+t] + e^{2t}(-2 + \mathrm{Cos}[x-t] - \mathrm{Sin}[x-t]) + \mathrm{Sin}(x+t)) + x(t + e^{-t} - 1)$$

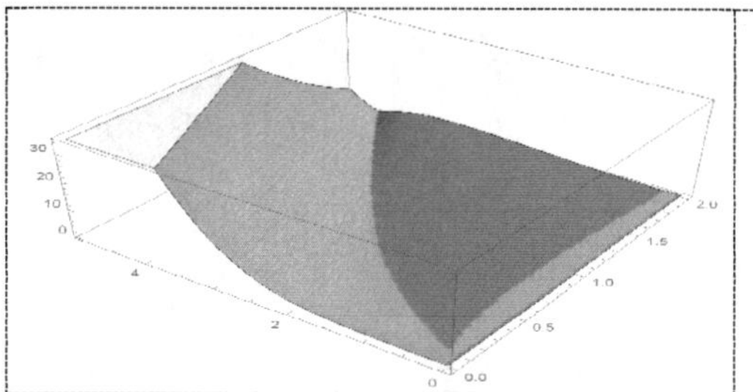

Figure 5.a. complex 3D plot for u, x and t.

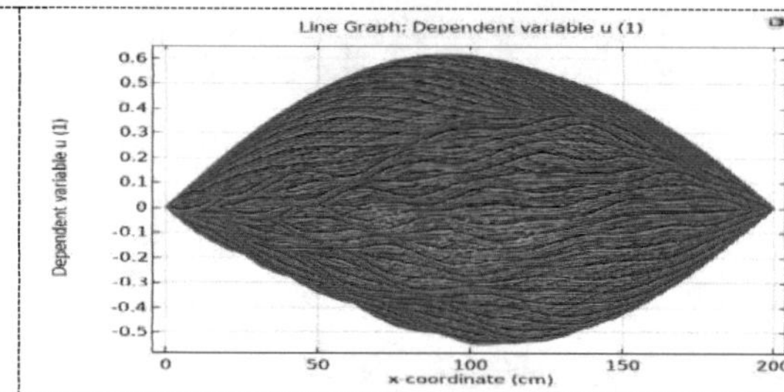

Figure 5.b. Real 2D animated plot for u and x.

2- a rod of length 3m placed on 2 supports at 0.5m and 2m with

Initial position function $u_{(x,0)} = p(x) = 0$, *Damping of* $h^*(x,t) = xe^{-t}$

Initial velocity function $u_{t_{(x,0)}} = v(x) = (1 - \cos(x))e^{-x}$, $c = 1$

$u_{tt} - u_{xx} = xe^{-t}$, *applying the previous formula* $u_{xx} - u_{yy} = -xe^{-y}$

$u_{(x,y)} = \frac{1}{2}(P(x-y) + P(x+y)) + \frac{1}{2}\int_{x_o-y_o}^{x_o+y_o} v(x)\, dx - \frac{1}{2}\int_0^{y_o}\int_{x_o-(y_o-y)}^{x_o+(y_o-y)} h(x,y) dxdy$

$u_{(x,y)} = \frac{1}{2}\int_{x_o-y_o}^{x_o+y_o}(1-\cos(x))e^{-x}\,dx + \frac{1}{2}\int_0^{y_o}\int_{x_o-(y_o-y)}^{x_o+(y_o-y)} xe^{-y}\,dx\,dy$ as shown in fig. (6)

$$u = \frac{e^{-x-t}}{4}(2 - \mathrm{Cos}[x+t] + e^{2t}(-2 + \mathrm{Cos}[x-t] - \mathrm{Sin}[x-t]) + \mathrm{Sin}(x+t)) + x(t + e^{-t} - 1)$$

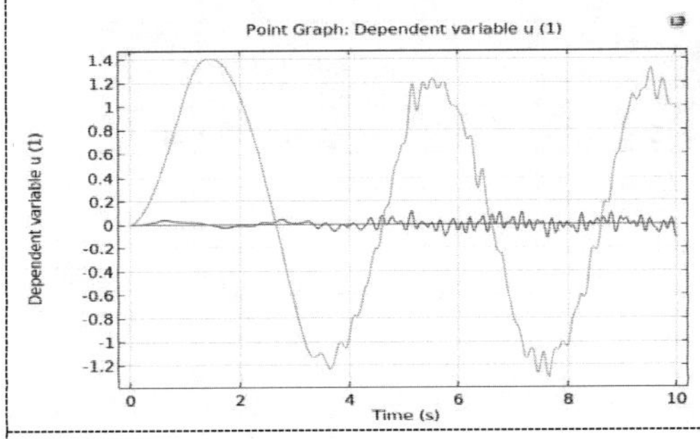

Figure 6.a. Real 2D animated plot for u and t.

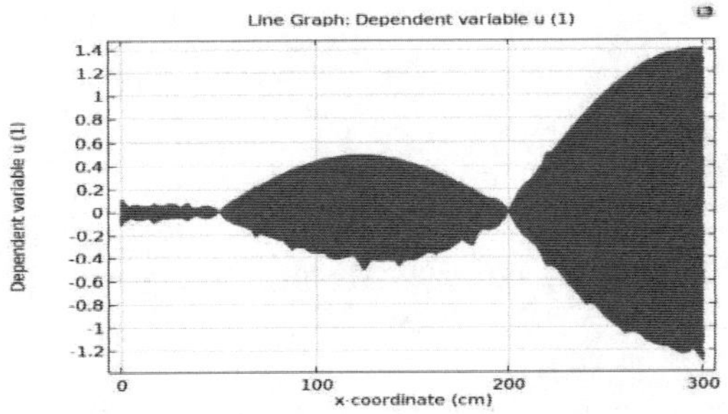

Figure 6.b. Real 2D animated plot for u and x.

4. Elliptical equations

4.1. Laplace equation

$u_{xx} + u_{yy} = 0 \quad A = 1 \quad B = 0 \quad C = 1 \qquad B^2 - 4AC = -4 < 0$

$A\lambda^2 - B\lambda + c = 0 \quad \lambda^2 + 1 = 0 \quad \lambda = \pm i \qquad \frac{dy}{dx} = i \qquad \frac{dy}{dx} = -i$

$\int dy = \int i\,dx \qquad iy = -x + c \qquad x + iy = c1 \quad let\ \varepsilon = x + iy$

$\int dy = \int -i\,dx \qquad iy = x + c \qquad x - iy = c2 \quad let\ \eta = x - iy$

$u_{xx} = u_{\varepsilon\varepsilon}{\varepsilon_x}^2 + 2u_{\varepsilon\eta}\varepsilon_x\eta_x + u_{\eta\eta}{\eta_x}^2 + u_\varepsilon\varepsilon_{xx} + u_\eta\eta_{xx} = -u_{\varepsilon\varepsilon} + 2u_{\varepsilon\eta} - u_{\eta\eta}$

$u_{yy} = u_{\varepsilon\varepsilon}{\varepsilon_y}^2 + 2u_{\varepsilon\eta}\varepsilon_y\eta_y + u_{\eta\eta}{\eta_y}^2 + u_\varepsilon\varepsilon_{yy} + u_\eta\eta_{yy} = u_{\varepsilon\varepsilon} + 2u_{\varepsilon\eta} + u_{\eta\eta}$

$u_{xx} + u_{yy} = 4u_{\varepsilon\eta} \quad so\ the\ canonical\ form\ is \qquad 4u_{\varepsilon\eta} = 0$

$u_\varepsilon = g(\varepsilon)\ ,\ u = f(\varepsilon) + g(\eta) = f(x+iy) + g(x-iy) = f(z) + g(\bar{z}) = u_{(x,y)} + iv_{(x,y)}$

General method for particular solution $u = u_{(x,y)} + iv_{(x,y)}$ u, v should be analytic and harmonic function $u_x = v_y\ ,\ u_y = -v_x$ *where* u, v are Real function of Real variables then the Real and Imaginary part of u each represents a solution for Laplace P.D.E or any combination of them

As Laplace equation is symmetric so the solution should be radial so we can set $u = v(r)$

$r = \sqrt{x^2+y^2}\ ,\quad u_x = v_r r_x = \frac{x}{r}v_r\ ,\quad u_{xx} = \frac{v_r}{r} + \frac{x^2}{r^2}v_{rr} - \frac{x^2}{r^3}v_r$

$u_y = v_r r_y = \frac{y}{r}v_r\ ,\quad u_{xx} = \frac{v_r}{r} + \frac{y^2}{r^2}v_{rr} - \frac{y^2}{r^3}v_r\ ,\quad u_{xx} + u_{yy} = \frac{1}{r}v_r + v_{rr}$

By this method the P.D.E reduced into ode where $\frac{1}{r}v_r + v_{rr} = 0$ solving for v by integration

$v = c\ln\left(\frac{1}{r}\right) + c^* \quad which\ is\ the\ fundmental\ sol.where \quad c = \frac{1}{\pi}\ ,\ c^* = 0$

$u = v = \frac{-1}{2\pi}\ln(x^2+y^2) = \frac{-1}{2\pi}(\ln(x+iy) + \ln(x-iy))$ *from general solution* shown in fig. (7)

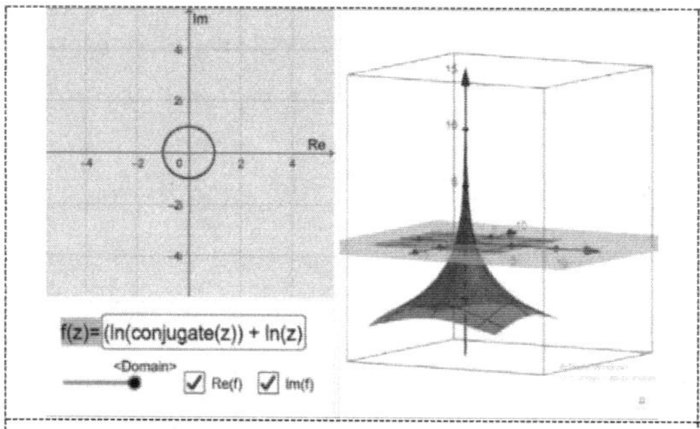

Figure 7.a. complex 3D plot for u, x and y.

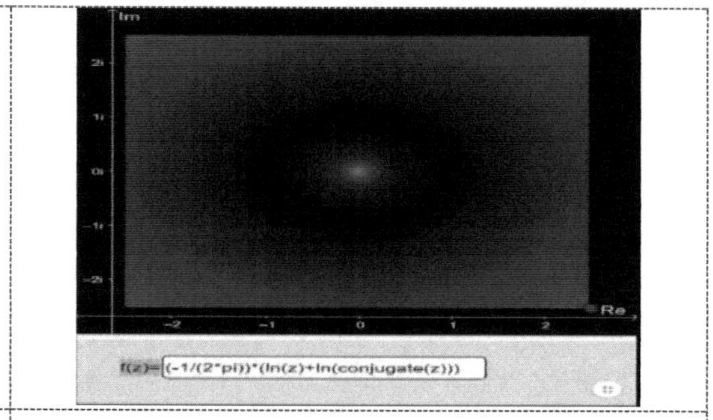

Figure 7.b. complex contour plot for x and y.

4.1.1. Physical application

1-Electrostatic potential charge in free region where the potential in the rectangle whose upper side is kept at potential 110 V and whose other sides are grounded.

$0 \leq x \leq 40 \ , \ 0 \leq y \leq 20 \ , \ la\ place\ equ. \ \ u_{xx} + u_{yy} = 0\ (cartesian)$

where u is the potential as shown fig. (8)

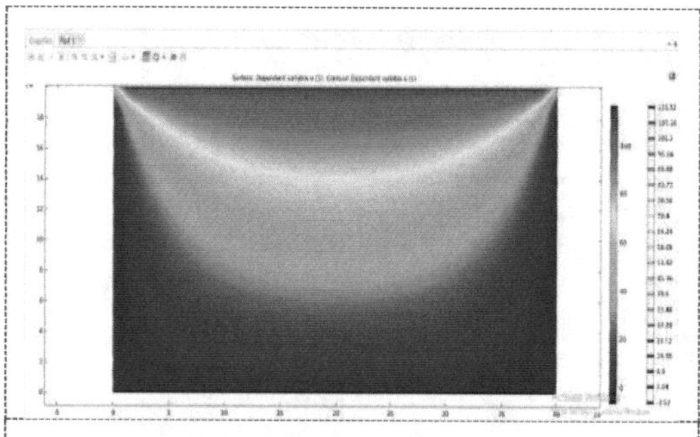

Figure 8.a. Real 2D plot for u and x.

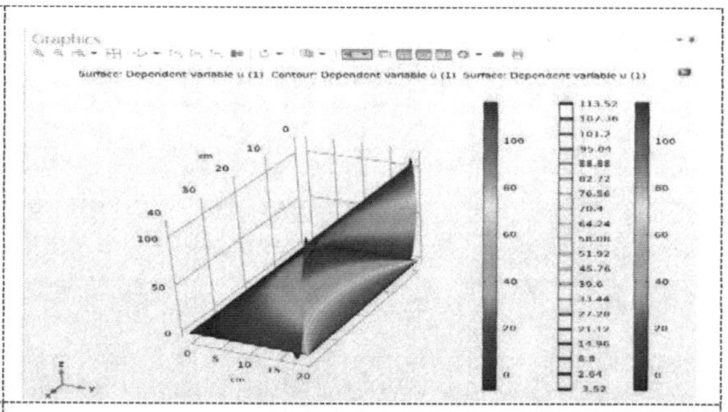

Figure 8.b. Real 3D plot for u, y and x.

2-The potential flow of an ideal incompressible fluid about a circular cylinder of radius R with a constant incident velocity v $\ la\ place\ equation \ \ u_{rr} + \frac{1}{r}\, u_r + \frac{1}{r^2}\, u_{\theta\theta} = 0 \ \ (cylinderical)$

$u = f(\varepsilon) + g(\eta) = f(\ln(r) - i\theta) + g(\ln(r) + i\theta) = Ar^n e^{in\theta} + \frac{B}{r^n} e^{-in\theta}$

$Re = \left(Ar^n + \frac{B}{r^n}\right)\cos(n\theta) \ , \ Im = \left(Ar^n - \frac{B}{r^n}\right) sin(n\theta) \quad by\ multiplying\ Re, Im\ we\ get$

$u = \left(Ar^n + \frac{B}{r^n}\right)(\mathrm{Ccos}(n\theta) + \mathrm{Dsin}(n\theta)) \qquad n = 1,2,3,4,5, \ldots$

We are going to solve this P.D.E twice with different initial and boundaries once for stream lines Then for velocity potential.

$u_{(R,\theta)} = 0 \ , \ \ r = R \ \ , \ \ r \to \infty \ , \ u \to vrsin(\theta)\ from\ IC, BC$

$Ar^n(\mathrm{Ccos}(n\theta) + \mathrm{Dsin}(n\theta)) = vrsin(\theta) \quad comparing\ coffecient \quad n = 1 \ , c = 0 \ , AD = v$

$\left(AR + \frac{B}{R}\right)(\mathrm{Dsin}(\theta)) = 0 \ , \ B = -AR^2$, substitute $in\ original\ so$

$u = \left(Ar^n + \frac{B}{r^n}\right)(\mathrm{Ccos}(n\theta) + \mathrm{Dsin}(n\theta)) = \left(Ar - \frac{AR^2}{r}\right) Dsin\theta$

$u = v\left(r - \frac{R^2}{r}\right)\sin(\theta) \quad where\ u\ is\ stream\ function$ as shown in fig. (9)

Solving the same P.D.E again for velocity potential where;

$u_{(R,\theta)} = 2vRcos(\theta) \quad , \quad r = R \quad , \quad r \to \infty \ , \ u \to vrcos(\theta)$
$comparing\ coffecients\ to\ get\ D = 0\ ,\ n = 1\ ,\ AC = v\ ,\ B = AR^2\ then\ substute$
$u_{(r,\theta)} = v\left(r + \frac{R^2}{r}\right)\cos(\theta) \quad where\ u\ is\ velocity\ potential$ as shown in fig. (10)

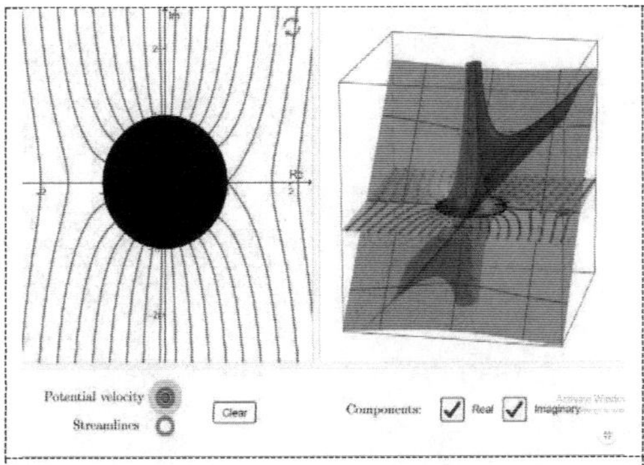

Figure 9. complex 3d plot and 2d plot for velocity potential.

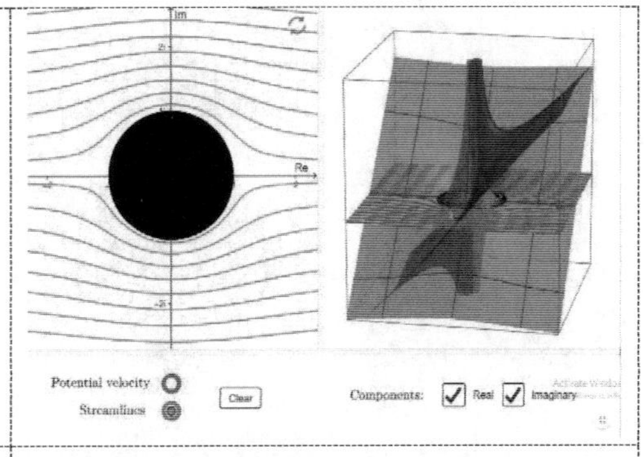

Figure 10. complex 3d plot for and 2d plot for streamlines.

By adding stream lines and velocity potential to get the potential flow
$U = v\left(r - \frac{R^2}{r}\right)\sin(\theta) + v\left(r + \frac{R^2}{r}\right)\cos(\theta)$ as shown in fig. (11)

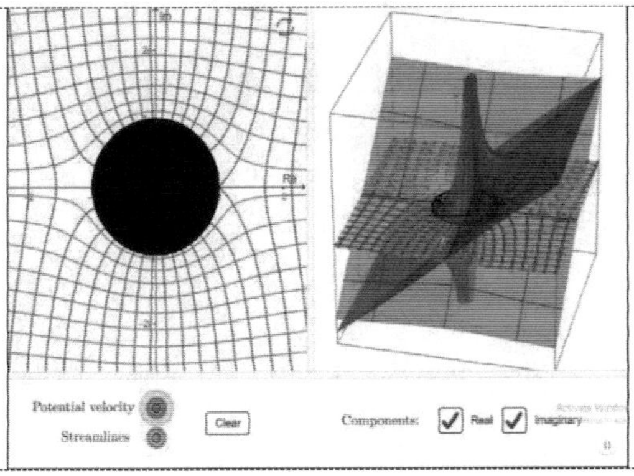

Figure 11.a. complex 3d plot for u and 2d plot for stream lines and velocity potential

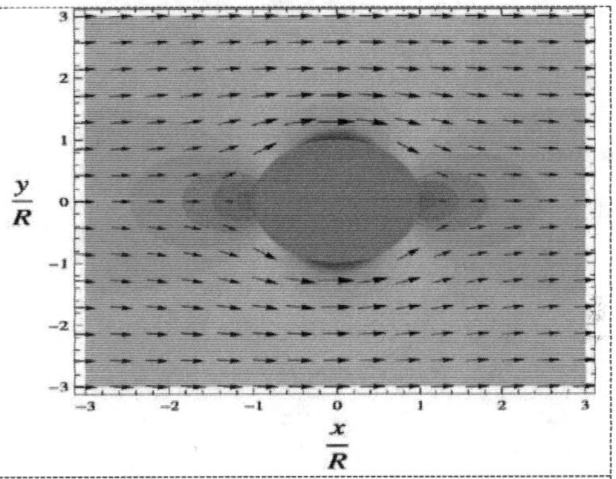

Figure 11.b. animated vector field plot for circular cylinder.

5. Parabolic equation

5.1. Diffusion equation
$u_t = ku_{xx} \quad , \quad |x| < \infty \quad , \quad t > 0$
$let\ a\ solution\ of \quad t = x^2$ satisfying the P.D.E where $\frac{x}{\sqrt{t}} = 1 \qquad let \quad \varepsilon = \frac{x}{\sqrt{t}}$
suppose $u = t^{-\alpha} f\left(\frac{x}{\sqrt{t}}\right)$ such that the total energy is conserved and preserved
$u_t = \frac{-t^{-\alpha-1}}{2}\varepsilon f^{\backslash}(\varepsilon) - \alpha t^{-\alpha-1} f(\varepsilon) \quad , \quad u_{xx} = t^{-\alpha-1} f^{\backslash\backslash}(\varepsilon)$ substitute in P.D.E
$kf^{\backslash\backslash}(\varepsilon) + \frac{\varepsilon}{2} f^{\backslash}(\varepsilon) + \alpha f(\varepsilon) = 0 \ , \quad take \ \ \alpha = \frac{1}{2} \quad , \quad kf^{\backslash\backslash}(\varepsilon) + \frac{1}{2}\left(\varepsilon f^{\backslash}(\varepsilon) + f(\varepsilon)\right) = 0$
$kf^{\backslash\backslash}(\varepsilon) + \frac{1}{2}\left(\varepsilon f(\varepsilon)\right)^{\backslash} = 0 \quad , \quad \left(kf^{\backslash}(\varepsilon) + \frac{1}{2}\varepsilon f(\varepsilon)\right)^{\backslash} = 0 \quad by\ integration\ we\ get$

$kf^{\backslash}(\varepsilon)+\frac{1}{2}\varepsilon f(\varepsilon)=c$ where this O.D.E has infinite solutions, we take $c=0$

$kf^{\backslash}(\varepsilon)+\frac{1}{2}\varepsilon f(\varepsilon)=0$ which is called fundamental solution by integration again

$f(\varepsilon)=Ae^{-\frac{\varepsilon^2}{4kt}}$ where A is integration constant can be determined by

$\because\ total\ energy\ is\ conserved \qquad \therefore \int_{-\infty}^{\infty}u(x,t)\,dx=1 \quad , \quad A=\frac{1}{\sqrt{4\pi k}}$

$u=t^{-\alpha}f(\varepsilon)=\frac{1}{\sqrt{4\pi kt}}e^{-\frac{x^2}{4kt}} \qquad which\ is\ a\ fundemental\ solution\ of\ pde$

$let\ k=\frac{1}{4} \quad , \quad u=\frac{1}{\sqrt{\pi t}}e^{-\frac{x^2}{t}}$ as shown in fig. (12)

5.1.1. *Physical application*

1- low frequency AC submarine cable where the cable is made such that inductance L and leakage of conductance G are neglected where; general telegraph equation

$i_{xx}=LCi_{tt}+(RC+GL)i_t+RGi \quad , \quad l=G=0 \quad , R\ ressistance\ , \quad C\ capictance$

$i_{tt}-\frac{1}{RC}\,i_{xx}=0$ low freq. AC similar to heat equation

$i(x,t)=\frac{1}{\sqrt{\frac{4\pi t}{RC}}}e^{\frac{\frac{-x^2}{4t}}{RC}} \ , \ R=\frac{4}{C}\ , \ i(x,t)=\frac{1}{\sqrt{\pi t}}e^{-\frac{x^2}{t}}$ as shown in fig. (13)

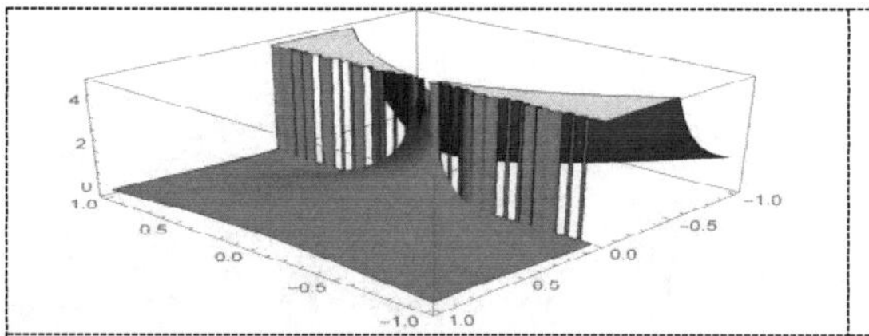

Figure 12. complex 3D plot for u, x and t.

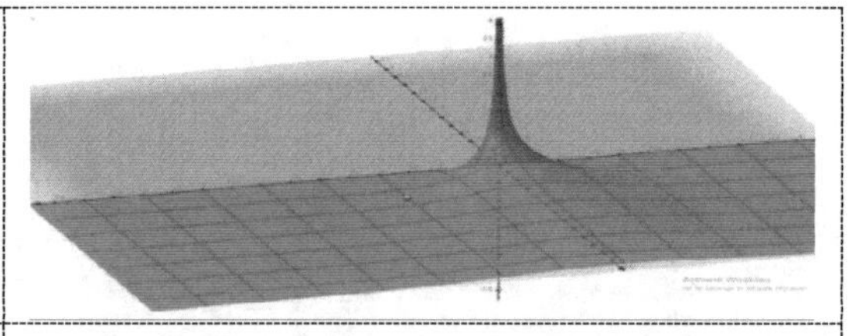

Figure 13. Real 3d plot for u, x and t.

6. Conclusion

The second-order linear PDEs can be classified into three types, which are invariant under changes of variables. The types are determined by discriminant. This exactly corresponds to the different cases for the quadratic equation satisfied by the slope of the characteristic curves. Hyperbolic equations have two distinct families of (real) characteristic curves, parabolic equations have a single family of characteristic curves, and the elliptic equations have none. All the three types of equations can be reduced to canonical forms to be simulated and modeled allowing the analysis of physical phenomena to predict the variance over time.

References

[1] Abul Hasan Siddiqi, & Kuzman Adzievski. (2014). *Introduction to partial differential equations for scientists and engineers using Mathematica.* 222 Rosewood Drive, Danvers: Taylor & Francis Group, LLC CRC Press.

[2] Andrei D. Polyanin, & Vladimir E. Nazaikinskii. (2016). *Handbook of linear partial differential equations for engineers and scientists* (2nd ed.). 222 Rosewood Drive, Danvers: Taylor & Francis Group, LLC CRC press.

[3] D. Gilbarg, & N. S. Trudinger. (1983). *Elliptic Partial Differential Equations* (Vol. 224). Grundlehren: Springer-Verlag, Berlin.

[4] ERWIN KREYSZIG. (2011). *ADVANCED ENGINEERING MATHEMATICS* (10th edition ed.). United States of America: Laurie Rosatone John Wiley & Sons, Inc.

[5] F. John. (1982). *Partial Differential Equations.* New York: Springer-Verlag.

[6] Lawrence c.Evans. (2010). *Partial Differential Equation* (2nd ed., Vol. 19). 201 Charles Street, Providance, Rhode Island: american mathematical society.

[7] S. L. SOBOLEV, & A. J. Lohwater. (1964). *PARTIAL DIFFERENTIAL EQUATIONS OF MATHEMATICAL PHYSICS* (3rd ed.). Moscow: Pergamon Press.

[8] Stanley J. Farlow. (1993). *Partial Differential Equations for Scientists and Engineers.* (Dover, Ed.) John Wiley & Sons Ltd.

[9] Tyn Myint, & Lokenath Debnath. (2007). *Linear Partial Differential Equations for Scientists and Engineers* (4th ed.). USA: (Birkhauser Boston, c/o Springer Science ¨ +Business Media LLC, 233Spring Street, New York, NY 10013, USA.

[10] Walter A. Strauss. (2008). *Partial differential* (second edition ed.). Chichester: John Wiley & Sons Ltd.

Analytical solution of diffusion equation under chemical reaction and wet deposition from line Source

Khaled S M Essa, Ahmed M Mosallem and H M Taha*

*The corresponding author

Department of Mathematics and theoretical physics, Nuclear Research Centre, Egyptian Atomic Energy Authority, Cairo, Egypt

E-mail: hanaataha3@yahoo.com

Abstract.The pollutants from the environment are absorbed by wet deposition which increases the pollutants transport. To create and produce different substances, the constituent atoms of the reactants should be rearranged by the chemical reaction. The introduced theoretical model demonstrate the dispersal of pollutants inherited from point and line source without and with both wet deposition and chemical reaction, respectively. The advection-diffusion equation is solved. Taking the wind speed is a function of downwind distance, lateral eddy diffusivity is a function of wind speed and vertical eddy diffusivity is a function of both wind speed and vertical height of power "z". In this paper the variable separable method and Hankel transform are used to get the concentration from point and line source. Predicted concentration of pollutants in unstable condition is done. Variation concentrations of both observed and predicted of radioactive Iodine135 (I-135), inside Research Reactor at Inshas Egyptian Atomic Energy Authority, in unstable condition is shown without chemical reaction and wet deposition. It is discovered that the proposed model is a good harmony with observed concentrations data and the proposed data inside a factor of two. Also, studying the effect of wet deposition and chemical reaction on concentration of air contaminants of line source is analysed. Keywords: The equation of advection diffusion; separable method; Hankel transform; Point and line source.

1. Introduction

Atmospheric concentration effects on human health by deposition of air contaminants. It was very supported for emission contaminants and their surroundings in urban areas in developed and developing countries. Several studies have been done for studying the effect of air contaminants on human health. Studying the winds isn't adequate to forecast atmospheric pollutants in large city region scale [1]. The influence of eddy diffusivity could be made variations in advection diffusion-equation, which has been studied [2]. Two-dimensional models analytically using two types of eddy diffusivities was solved [3]. The effect of wind, chemical reaction and wet deposition in an urban area were not considered by those two models. The model for the dispersal of pollutants inherited from line source with wind considered, is estimated in urban heat island [4]. A flux-Profile relationship in atmospheric surface layer has been given [5]. A numerical model of advection-diffusion, with time-dependent due to area source under chemical reaction, was estimated [6]. Gravitational settling velocity and leakage velocity of contaminants, that not consider wet deposition, have been studied [7]. Chemical reactive of pollutants with different shapes of wind in area source using numerical model has been studied [8]. Atmospheric diffusion equation using longitudinal wind speed depending on source distance was investigated [9]. Analytical of air pollution emitted from line source under chemical reaction and wet deposition has been considered [10].

In this work, an analytical model for the dispersal of pollutants inherited from line source, with wet deposition and chemical reaction, is established. Taking the wind speed is a function of downwind distance, lateral eddy diffusivities is a function of wind speed and vertical eddy diffusivities is a function of both wind speed and vertical height of power z. A separation method and Hankel transform are used to solve above model.

2. Mathematical Treatment

The partial differential equation describing advection-diffusion equation can be written as:

$$u\frac{\partial C(x,y,z)}{\partial x}=\frac{\partial}{\partial y}\left(k_y\frac{\partial C(x,y,z)}{\partial y}\right)+\frac{\partial}{\partial z}\left(k_z\frac{\partial C(x,y,z)}{\partial z}\right)-k_{cw}C(x,y,z) \tag{1}$$

where, $C(x,y,z)$ is the pollutants concentration (Bq/m^3), k_y and k_z are the eddy diffusivities in y and z directions respectively, u is as function of downwind distance, x is downwind distance, $k_{cw}=k_c+k_w$ where,k_c, k_w are the chemical reaction and wet deposition coefficients respectively.

Therefore, solving this problem using the technique of separation variable, assuming that:

$$u(x)=k_0u_*(1-ax) \tag{2a}$$

$$k_y(x)=\beta u \tag{2b}$$

$$k_z(x,z)=\gamma u z^n, \qquad z\neq 0 \tag{2c}$$

where, $\beta=0.16\left(\frac{\mathrm{W}*}{u}\right)^2$ and $\gamma=0.31\left(\frac{\mathrm{W}*}{u}\right)^2$ are turbulence constants, W $*$ is a convective vertical velocity and 'n' is a parameter depends on stability conditions [11]. Then Eq. (1) can be written as:

$$\frac{\partial C}{\partial x}=\beta\frac{\partial^2 C}{\partial y^2}+\gamma\frac{\partial}{\partial z}\left(z^n\frac{\partial C}{\partial z}\right)-\frac{k_{cw}}{k_0u_*(1-ax)}C \tag{3}$$

Eq. (3) is solved with the following boundary conditions:

(a) Flux at vertical height vanishes at ground level and at height of boundary layer i.e.

$$k_z\frac{\partial C}{\partial z}=0, \qquad \text{at } z=0,h \tag{4a}$$

(b) Flux at crosswind direction vanishes at centre line and at height Ly i.e.

$$k_y\frac{\partial C}{\partial y}=0, \qquad at y=0,L_y \tag{4b}$$

© Mass continuity is applied as follows:

$$uC(x,y,z)=Q\delta(y)\delta(z-h_s) \qquad \text{at } x=0 \tag{4c}$$

where, Q is the emission rate.

(d)The concentration tends to zero as follows:

$$C(x,y,z)\to 0 \qquad \text{as } y\to\pm\infty \quad and z\to\infty \tag{4d}$$

where, the mixing height (m), a large distance in the y direction, a stack height, and a Dirac delta function are mentioned by h, Ly, hs and δ respectively. The general solution of Eq. (3) is assumed by the following form:

$$C(x,y,z)=\varphi(x,y)\psi(x,z) \tag{5}$$

Using Eq. (5), then Eq. (3) could be converted into the following form

$$\frac{1}{\varphi}\frac{\partial\varphi}{\partial x}+\frac{1}{\psi}\frac{\partial\psi}{\partial x}=\frac{\beta}{\varphi}\frac{\partial^2\varphi}{\partial y^2}+\frac{\gamma}{\psi}\frac{\partial}{\partial z}\left(z^n\frac{\partial\psi}{\partial z}\right)-\frac{k_{cw}}{k_0u_*(1-ax)} \tag{6}$$

Subdividing Eq. (6) into the following two equations as follows:

$$\frac{1}{\varphi}\frac{\partial\varphi}{\partial x}=\frac{\beta}{\varphi}\frac{\partial^2\varphi}{\partial y^2}-\frac{k_{cw}}{k_0u_*(1-ax)} \tag{7a}$$

$$\frac{1}{\psi}\frac{\partial\psi}{\partial x}=\frac{\gamma}{\psi}\frac{\partial}{\partial z}\left(z^n\frac{\partial\psi}{\partial z}\right) \tag{7b}$$

2.1 First Description Model

Using the method of separation variable to solve Eq. (7a), so assuming the following form

$$\varphi(x,y)=\chi_l(x)\eta_l(y) \tag{8}$$

By substituting Eq. (8) into Eq. (7a) then, one gets:

$$\frac{1}{\chi_l(x)}\frac{\partial\chi_l(x)}{\partial x}+\frac{k_{cw}}{k_0u_*(1-ax)}=\frac{\beta}{\eta_l(y)}\frac{\partial^2\eta_l(y)}{\partial y^2}=-\lambda_l^2 \tag{9}$$

where, $\lambda_l$is a separation constant. Then, we have two differential equations as follows:

$$\frac{\partial\chi_l}{\partial x}=-\frac{k_{cw}}{k_0u_*(1-ax)}\chi_l-\lambda_l^2\chi_l \tag{10a}$$

$$\frac{\partial^2\eta_l}{\partial y^2}=\frac{-\lambda_l^2}{\beta}\eta_l \tag{10b}$$

The solution of Eqns. (10 a, and b) has the form:

$$\chi_l(x)=c_1(1-ax)^{\frac{k_{cw}}{ak_0u_*}}e^{-\lambda_l^2x} \tag{11a}$$

$$\eta_l(y)=c_2cos\left(\frac{\lambda_l}{\sqrt{\beta}}y\right)+c_3sin\left(\frac{\lambda_l}{\sqrt{\beta}}y\right) \tag{11b}$$

where, the constants $c_1, c_2,$ and c_3 are calculated by using the boundary, then $c_3=0$ and $\lambda_l=\frac{l\pi\sqrt{\beta}}{L_y}$, $l=$ 0,1,2,...

Then the solution of Eq. (10) is as follows:

$$\varphi(x,y)=\sum_{l=0}^{\infty}B_l(1-ax)^{\frac{k_{cw}}{ak_0u_*}}e^{-\lambda_l^2x}cos\left(\frac{l\pi}{L_y}y\right) \tag{12}$$

where,$B_l=c_1c_2$, Using the boundary condition Eq. (4c) and Eq. (5), one gets: $B_0=\frac{1}{L_y}$, $B_i=\frac{2}{L_y}$, $l=$ 1,2,3, ...

2.2 Second Description Model

Eq. (7b) is written in the form:

$$z^n\frac{\partial^2\psi}{\partial z^2}+nz^{n-1}\frac{\partial\psi}{\partial z}-\frac{1}{\gamma}\frac{\partial\psi}{\partial x}=0 \tag{13}$$

Following by taking substitution $\rho=z^{\frac{2-n}{2}}$ [12], then Eq. (13) will be written as follows:

$$\frac{\partial^2 \psi}{\partial \rho^2} + \frac{n}{2-n}\rho^{-1}\frac{\partial \psi}{\partial \rho} - \frac{1}{\gamma}\left(\frac{2}{2-n}\right)^2 \frac{\partial \psi}{\partial x} = 0 \quad (14)$$

Substituting by $\psi(x,\rho) = \rho^m \omega(x,\rho)$, $m = \frac{1-n}{2-n}$[12]. Then Eq. (14) will be

$$\frac{\partial^2 \omega}{\partial \rho^2} + \frac{1}{\rho}\frac{\partial \omega}{\partial \rho} - \frac{m^2}{\rho^2}\omega - \frac{4(1-m)^2}{\gamma}\frac{\partial \omega}{\partial x} = 0 \quad (15)$$

Now by applying Hankel transformation [12] which was defined by

$$\mathcal{H}_m\{f(z)\} = \tilde{f}(s) \equiv \int_0^\infty f(z) J_m(sz) z dz$$

Such that the Bessel differential operator and its Hankel transformation [12] also were defined by

$$\Delta_m f(z) \equiv \frac{d^2 f(z)}{dz^2} + \frac{1}{z}\frac{df(z)}{dz} - \left(\frac{m}{z}\right)^2 f(z)$$

$$\mathcal{H}_m\{\Delta_m f(z)\} \equiv -s^2 \tilde{f}(s)$$

Then Eq. (15), after applying Hankel transformation [12], can be written as the following form

$$-s^2 \widetilde{\omega}(x,s) - \frac{4(1-m)^2}{\gamma}\frac{\partial \widetilde{\omega}}{\partial x} = 0 \quad (16)$$

which, has the solution.

$$\widetilde{\omega}(x,s) = AExp\left[-\frac{\gamma x s^2}{4(1-m)^2}\right] \quad (17)$$

By using the condition Eq. (4c), the substitution $\rho = z^{\frac{2-n}{2}}$ and $\psi(x,\rho) = \rho^m \omega(x,\rho)$, $m = \frac{1-n}{2-n}$ then, one can get

$$A = \widetilde{\omega}(0,s) = \mathcal{H}_m\{\omega(0,\rho)\} = \frac{Q}{k_0 u_*}\int_0^\infty \rho^{-m}\delta(\rho^{2(1-m)} - h_s) J_m(s\rho)\rho d\rho$$

$$= \frac{Q h_s^{\frac{m}{2(1-m)}}}{2k_0 u_*(1-m)} J_m\left(s h_s^{\frac{1}{2(1-m)}}\right) \quad (18)$$

Then, Eq. (17) becomes:

$$\widetilde{\omega}(x,s) = \frac{Q h_s^{\frac{m}{2(1-m)}}}{2k_0 u_*(1-m)} J_m\left(s h_s^{\frac{1}{2(1-m)}}\right) Exp\left[-\frac{\gamma x s^2}{4(1-m)^2}\right] \quad (19)$$

Taking the inverse Hankel transformation [12] which was defined by

$$\mathcal{H}_m^{-1}\{\widetilde{\omega}(x,s)\} = \omega(x,\rho) \equiv \int_0^\infty \widetilde{\omega}(x,s) J_m(s\rho) s ds$$

Change the variable ρ into the variable z and the function $\omega(x,\rho)$ into the function $\psi(x,z)$

Then, Eq. (7b) has the solution

$$\psi(x,z) = \frac{Q(h_s z)^{\frac{1-n}{2}}}{(2-n)k_0 u_* \gamma x} Exp\left[-\frac{h_s{}^{2-n}+z^{2-n}}{(2-n)^2\gamma x}\right] I_{\frac{1-n}{2-n}}\left(\frac{2(h_s z)^{\frac{2-n}{2}}}{(2-n)^2\gamma x}\right) \quad (20)$$

Where I_m is the modified Bessel function of the first kind of order m.
By using Eq. (5), (12) and Eq. (20) then, Eq. (1) has the solution

$$C(x,y,z) = \frac{Q(h_s z)^{\frac{1-n}{2}}}{(2-n)k_0 u_* \gamma x} Exp\left[-\frac{h_s{}^{2-n}+z^{2-n}}{(2-n)^2\gamma x}\right] I_{\frac{1-n}{2-n}}\left(\frac{2(h_s z)^{\frac{2-n}{2}}}{(2-n)^2\gamma x}\right)$$

$$\times \sum_{l=0}^{\infty} B_l (1-ax)^{\frac{k_{cw}}{a k_0 u_*}} e^{-\lambda_l^2 x} \cos\left(\frac{l\pi}{L_y} y\right) \quad \mathbf{(21)}$$

2.3 Discussion and Results

The observed data of I-135 isotope concentration was obtained from dispersion experiments conducted in unstable condition air samples which was collecting around the Research Reactor at Inshas, Egyptian Atomic Energy Authority. The samples were collected at a height of 0.7 m above ground from a stack of height 43 m. The Reactor site was flat and controlled by soil of sand and a poor vegetation cover with a roughness length of 0.6 m. The power-law exponent ‘n’ values of eddy diffusivity as functions of air stability are considered [13] and given in Table (1). The measured concentration of I-135 isotope and the meteorological data during the experiments are considered [14] and given in Table (2). The predicted concentrations by Eq. (21) below the plume centre line are also given in Table (3). Variations between observed concentrations of radioactive I-135 in unstable condition at Inshas and predicted concentrations are shown in Figs. (1), (2) without chemical reaction and wet deposition (kcw=0). From these two figures, one finds that the proposed model is a good harmony with observed concentrations data and the proposed data inside a factor of two.

Table 1. Power-law exponent ‘n’ of eddy diffusivity as a function of air stability in urban area.

	A	B	C	D	E	F
n	0.85	0.85	0.80	0.75	0.60	0.40

Table 2. Meteorological data of the nine convective test runs at Inshas site in March and May 2006.

Run no.	Working hours of the source	Release rate (Bq)	Wind speed (m s^{-1})	Wind direction(deg)	W* (ms-1)	P-G stability class
1	48	1028571	4	301.1	2.27	A
2	49	1050000	4	278.7	3.05	A
3	1.5	42857.14	6	190.2	1.61	B
4	22	471428.6	4	197.9	1.23	C
5	23	492857.1	4	181.5	0.958	A
6	24	514285.7	4	347.3	1.3	D
7	28	1007143	4	330.8	1.51	C
8	48.7	1043571	4	187.6	1.64	C
9	48.25	1033929	4	141.7	2.1	A

Table 3. Observed and calculated concentrations for Run 9 experiments

Test	Downwind distance (m)	Observed conc. (Bq/m^3)	Predicted conc. Eq.(21) (Bq/m^3)
1	100	0.025	0.019
2	98	0.037	0.016
3	136	0.091	0.065
4	135	0.197	0.177
5	106	0.272	0.159
6	186	0.188	0.129
7	165	0.447	0.306
8	154	0.123	0.203
9	106	0.032	0.033

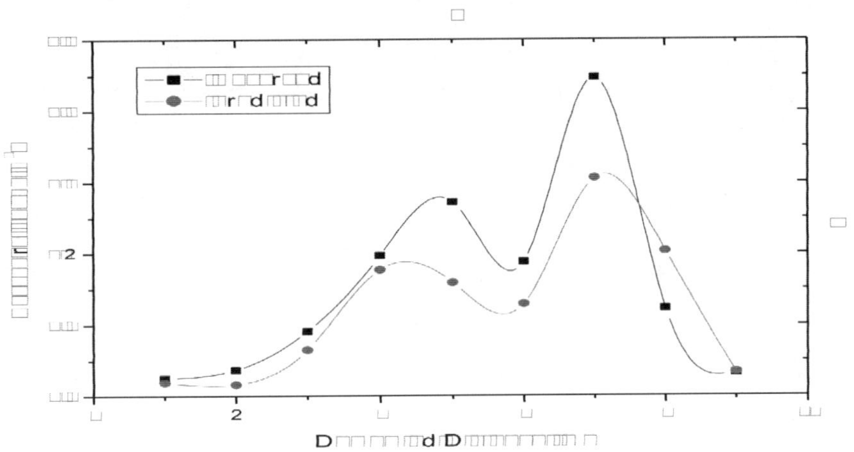

Figure 1. Variation of predicted and observed concentration (Bq/m^3) via downwind distance without chemical reaction and wet deposition

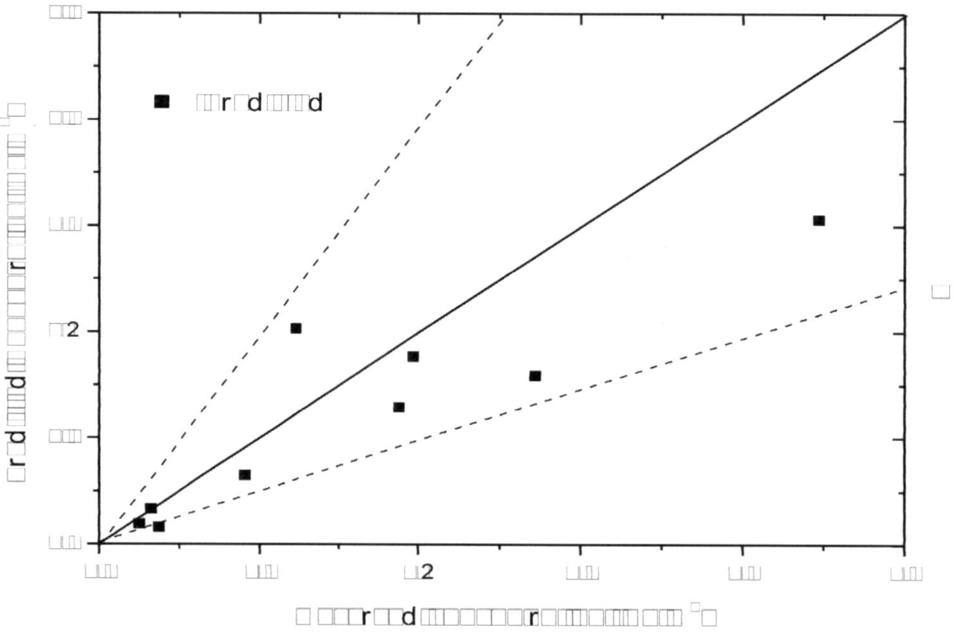

Figure 2. Diagram of Predicted model with corresponding observed concentrations, the solid line and dashed lines indicate to one to one and a factor of two respectively.

2.4 *Statistical Technique*

The statistical method between predicted and observed concentrationswasintroduced [15]. Where, NMSE is the normalized mean square error, FB is the Fraction Bias, COR is the Correlation Coefficient and FAC2 is the Factor of Two.

Table 4.The statistical method between predicted and observed concentration in unstable condition.

	NMSE	FB	COR	FAC2
Predicted	0.25	0.24	0.90	0.83

The statistical methods are recorded in table (4), which concluded that, the proposed model inside a factor of two with observed concentration data. Regarded to NMS, FB are nearly by zero and COR, FAC are nearly by one.

Also, in this work, the effect of chemical reaction and wet deposition on concentration of air contaminants of line source is considered. The concentration circulation is calculated at the city in unstable condition at downwind distance as follows $0 \leq x \leq 500m,$ taking stack height equals 0.5m, n=0.85, $K_{cw} = 0, 0.001, 0.01$ *and* 0.1, u=4.2m/s, Ly=100m and k0=0.4 and Q=3.2g/s and the vertical height “z” = 7, 8 and 10m.

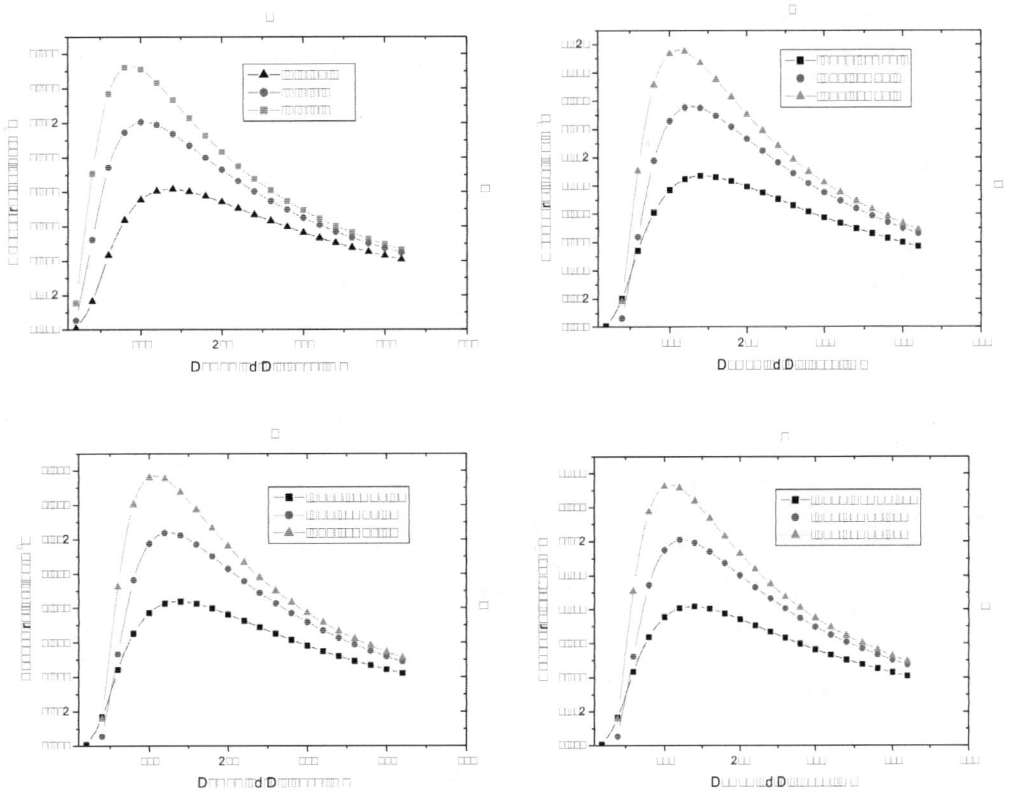

Figure3. The variation of the concentration with downwind distance at Kcw =0,0.1,0.01 and 0.001 respectively, where z =7, 8 and 10m.

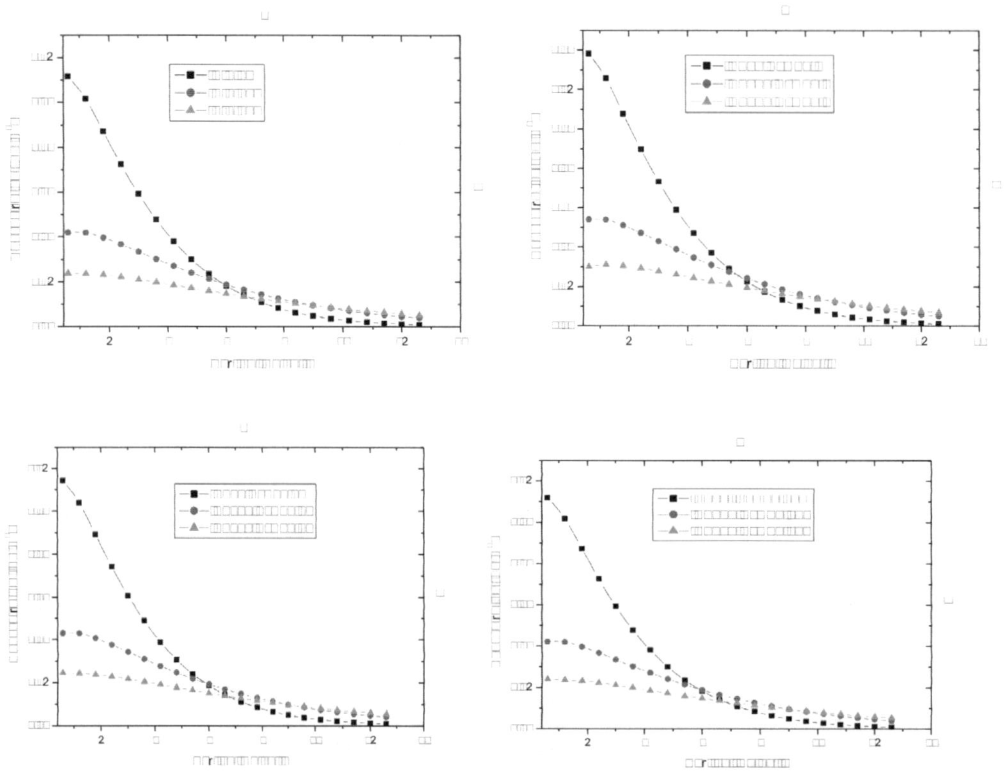

Figure 4. The variation of the concentration with vertical height (m) at Kcw =0,0.1,0.01 and 0.001 respectively, where x =50, 100 and 150m.

The concentration from point and line source via downwind distance (x) is shown in Fig. (3). At z=7m the concentration starts maximum value 0.016Bq/m3 with Kcw=0, and then increases to 0.020 Bq/m3 through line source at Kcw=0.1and after that the maximum concentration decreases to 0.11 Bq/m3 at Kcw=0.01 and 0.001. While at z=8m the concentration has maximum value 0.012Bq/m3 with Kcw=0 and then increases to 0.016Bq/m3 through line source at Kcw=0.1. Finally,at z=10m the concentration has maximum value 0.008Bq/m3 with Kcw=0 and then increases to 0.011Bq/m3 through line source at Kcw=0.1. For z=7m the maximum value of the concentration happens at x=100m at point source (kcw=0) but the maximum value of the concentration happensat x=130m at line source (kcw=0.1).
The concentration from point and line source via vertical height (z) is shown in Fig. (4).At downwind distance equals 50m the maximum value of the concentration has value 0.11 Bq/m3 at Kcw=0 from point source but the maximum value of the concentration equals 0.14 Bq/m3 at Kcw=0.1 which emissions from line source and after that the maximum concentration decreases to 0.11 Bq/m3 at Kcw=0.01 and 0.001. While the downwind distance increases to x=100 and 150m the maximum value of the concentration decreases. Also the concentration decreases to arrive zero when z=13m at x=50,100 and 150m.

3. Conclusions

A theoretical model is used to estimate concentration of air contaminants inherited from line source in unstable condition of atmosphere at Egyptian Atomic Energy Authority in existence of chemical reaction and wet deposition. The outcomes are expanded for the diffusion of air contaminants in the downwind distance and vertical height in unstable condition. This theoretical model is estimated using method of separable variable and Hankel transform. In nonexistence of chemical reaction and wet deposition, we find that the predicted concentration data and the observed concentration of Iodine 135 at Egyptian Atomic Energy Authority are inside a factor of two. Figures are plotted and the results are expanded for air contaminants concentration with and without chemical reaction and wet depositions for different heights (z = 7, 8 and 10m) and the downwind distance (x = 50, 100 and 100m) at the city in unstable condition. We notice that the concentration decreases whenever downwind distance increases. Also, we conclude that the concentration reduces with and without chemical reaction and wet deposition whenever downwind distance and vertical height increase. Maximum value of the concentration at z=7m and kcw=0.1(line source) and then decreases at z==8,10m at kcw=0.01and 0.001 (line source) respectively. The maximum value of the concentration from line source via vertical height (z) increases at Kcw=0.1 at downwind distance equals 50m and decreases at x=100 and 150m, at Kcw=0.01and 0.001 respectively.

References

[1] Griffiths J F 1970 Problems in urban air pollution in: AIAA 8th aerospace Science Meeting *(AIAA, New York)* 70-112.

[2] Marrouf A A, Essa K S M, E1- Otaify Maha S, Mohamed Adel S, Ismail Galal 2015 The Influence of Eddy diffusivity Variation on the Atmospheric Diffusion Equation *Open Journal of Air Pollution***04** 109-18.

[3] Essa K S M, Mina A N and Higazy M 2011 Analytical solution of diffusion equation in two dimensions using two forms of eddy diffusivities *Romanian journal of physics***56** 1228-40.

[4] Dilley J F, Yen K T 1971 Effect of mesoscale type wind on the pollutant distribution from a line source *Atmospheric Environment***5** 843-51.

[5] Businger J A, Wyngaard J C, Izumi Y and Bradley E F 1971 Flux Profile relationships in atmospheric surface layer *Journal al of the Atmospheric science***28** 181-9.

[6] Venktachalappa M, Sujith Kumar Khan and Khaleel Ahmed G Kakamari 2003 Time dependent mathematical model of air pollution due to area source with variable wind velocity and eddy diffusivity and chemical reaction *Proc Indian Nat. Sci Acad***69A No.6** 745-58.

[7] Suresha C M, Lakshminarayanachari K, Siddalinga Prasad M, and Pandurangappa C 2013 Numerical model of air pollutants emitted from an area source of primary and secondary pollutants with chemical reaction and gravitational settling with point sorceon the boundry *International Journal of Research in Environmental science and technology* 9-18.

[8] Lakshminarayanachari K, Sudheer Pai K L, Siddalinga Prasad M and Pandurangappa C 2013 A two-dimensional numerical model of primary pollutant emitted from an urban area source with mesoscale,*Dry deposition and chemical reaction Atmospheric Pollution Research***4** 106-16.

[9] Moreira D M and Taciana Toledo de Almeida Albuquerque 2016 Solution of the atmospheric diffusion equation with longitudinal wind speed depending on source distance. *RevistaBrasileira de Meteorologia***Vol 31No. 2** 202-10.

[10] Bhaskar C, LakshminarayanachariK and Pandurangappa C 2019 Analytical approach for air pollution emitted from line source with chemical reaction and wet deposition *International Journal of Research in Advent Technology*, **Vol.7 No.5**.

[11] Irwin J S 1979 A Theoretical variation of the wind profile power law exponent as a function of surface roughness and stability Atmos. *Environ***13** 191- 4

[12] Essa KSM, Ibrahim Mahmoud AE,Shalaby Ahmed S, Mosallem Ahmed M 2019 Analytical Solution of Two-Dimensional Diffusion Equation using Hankel Transform *Journal of Scientific and Engineering Research***6 No 6**6-11.

[13] Hanna S R, Briggs G A, Hosker R P Jr 1982 Handbook on atmospheric diffusion *Technical information center Department of energy United States*https://doi.org/10.2172/5591108.

[14] Essa K S M 2009Gaussian Plume Model Parameter for Ground Level and Elevated Sources Derived from the Atmospheric Diffusion Equation in the Neutral and Stable Conditions*Arab journal of nuclear sciences and application***Vol. 42 No. 3**.
[15] HannaSR 1989Confidence Limit for Air Quality Models as Estimated by Bootstrap andJacknife Resembling Methods*Atom. Environ***23**1385-95.

Journal of Physics: Conference Series 2304 (2022) 012005 doi:10.1088/1742-6596/2304/1/012005

Generalized thermoelastic in a fiber-reinforced medium with cylindrical cavity under G-N theory

S E Khader[1], A A Marrouf[1], M Khedr[2]

[1] Mathematics and Theoretical Physics Department, Nuclear Research Centre, Atomic Energy Authority, Cairo, Egypt
[2] University of Alexandria, Faculty of Science, Physics Department, Alexandria, Egypt

Email: a4_mmarrouf@yahoo.com

Abstract The present paper is aimed at studying the wave propagation for a problem of cylindrical cavity. The material is assumed to be fiber-reinforced with increased stiffness, strength and load bearing capacity. Using generalized thermoelasticity theory, Green and Nagdhi, with and without energy dissipation. Laplace transform techniques are used to derive the solution in the Laplace transform domain. The inversion process is carried out using a numerical method based on Fourier series expansions. The results obtained are displayed by graphs to clear the phenomena physical meaning and compeer with generalized thermoelasticity theories.

1. Introduction

The introduction of the fiber into the material mesh contributes to the improvement of the properties of the material. Reinforcement of the mesh is said to occur when the final composite material has properties in strength and flexibility that exceed that of the material alone. The extent of strength and elasticity that improves when reinforcing the material with fibers depends on the mechanical properties of both the backing fiber and the base of the material network, in addition to the ratio of their sizes to each other, as well as the length of the fiber, and how it is directed inside the material network. Fiber-reinforced plastics are used in a wide variety of applications in aerospace equipment, the automotive industry, and in engineering and military equipment. A continuum model is used to explain the mechanical properties of such materials. Craig and Hart [1] they derived the basic equations for the finite planar deformations of an ideal non-compressible fiber-reinforced material. Singh and Singh [2] studied the reflection of plane waves at the free surface of a fiber-reinforced elastic half-space. Bayones and Hussein [3] discussed the effects of thermal relaxation time and gravity of a two-dimensional problem with anisotropic semi-infinite solid media with the deformation of fiber-reinforced generalized thermoplastic. Sur et al. [4] have studied the thermoelastic interaction in a fiber-reinforced hollow cylinder due to thermal shock. Abbas and Othman [5] discussed the propagation of plane waves in fiber-reinforced, anisotropic thermoelastic half-space proposed by Lord-Shulman under hydrostatic initial stress. Abouelregal and Zenkour [6] have studied the generalized thermoelasticity problem for an infinite fiber-reinforced thick plate under initial stress. Many authors contributed to this subject [7-15].

The theory of generalized thermoelasticity with one relaxation time was introduced by Lord and Shulman [16]. In this theory Cattaneo-Maxwell law of heat conduction replaces the conventional

Fourier's law. The heat equation associated with this theory is a hyperbolic one and hence automatically eliminates the paradox of infinite speeds of propagation inherent in both the uncoupled and the coupled theories of thermoelasticity. For many problems involving steep heat gradients and when short time effects are sought this theory is indispensable. Sherief and El-Maghraby solved some crack problems for this theory [17-18]. Sherief and Hamza has obtained the solution of axisymmetric problems in spherical regions in [19] and in cylindrical regions in [20]. Sherief and Ezzat have obtained the solution in the form of series in [21]. Sherief and Dhaliwal used asymptotic expansions to obtain the solution of a 1D problem and to find the locations of the wave fronts and the speed of propagation of thermoelastic waves in [22]. This theory was extended to deal with micro polarity of the medium in [23]. Other works in the subject are [24-34].

2. Formulation of the problem

Let (r, φ, z) be cylindrical polar coordinates, with the z-axis as the axis of the cylinder. Consider a fiber-reinforced thermoelastic infinite body having a cylindrical cavity occupying the region. The surface of the cavity is subjected to a known temperature shock. The surface of the cavity is assumed to be traction free.

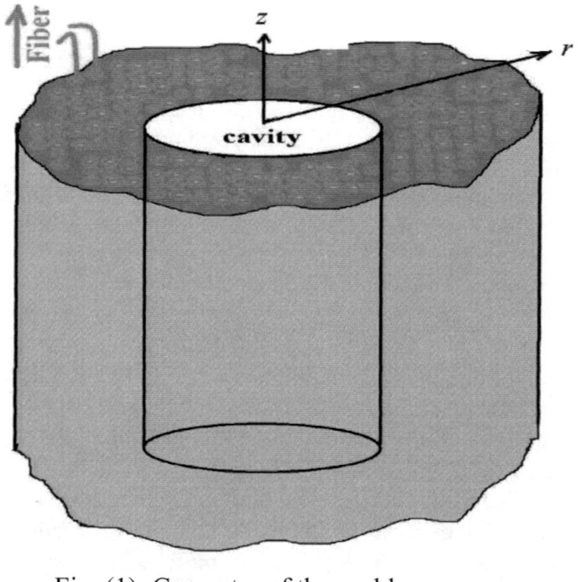

Fig. (1): Geometry of the problem

The displacement u has components

$$\mathbf{u} = (u(r,t),0,0)$$

This displacement field will produce the following strain components

$$e_{rr} = \frac{\partial u}{\partial r}, \quad e_{\phi\phi} = \frac{u}{r}, \quad e_{zz} = e_{rz} = e_{z\phi} = e_{r\phi} = 0 \tag{1}$$

The cubic dilatation e is thus given by

$$e = \frac{\partial u}{\partial r} + \frac{u}{r} = \frac{1}{r}\frac{\partial (ru)}{\partial r} \tag{2}$$

The components of the thermoelastic stress tensor σ_{ij} are given by [4]

$$\sigma_{ij} = \lambda e_{kk}\delta_{ij} + 2\mu_T e_{ij} + \alpha\left(a_k a_m e_{km}\delta_{ij} + a_i a_j e_{kk}\right) + 2(\mu_L - \mu_T)\left(a_i a_k e_{kj} + a_j a_k e_{ki}\right)$$
$$+ \beta a_k a_m e_{km} a_i a_j - \beta_{ij}(T - T_0)$$

Where e_{ij} are the strain components, λ, μ_T are the elastic constants, $\alpha, \beta, \mu_L, \mu_T$ are shear moduli for shear on parallel planes to the fibers, δ_{ij} are the Kronecker delta, $\beta_{rr} = \beta_{\phi\phi} = \beta_{zz} = \gamma$ and, $\gamma = (3\lambda + 2\mu_T)\alpha_t$, α_t is the thermal expansion coefficient. T is the temperature over the reference temperature T_0 and the fiber direction is $a \equiv (a_1, a_2, a_3) \equiv (0,0,1)$, where $a_1^2 + a_2^2 + a_3^2 = 1$. Now the constitutive relations for the present case are

$$\sigma_{rr} = (\lambda + 2\alpha + 4\mu_L - 2\mu_T + \beta)\frac{\partial u}{\partial r} + (\lambda + \alpha)\frac{u}{r} - \gamma(T - T_0) \tag{3a}$$

$$\sigma_{\phi\phi} = (\lambda + 2\mu_T)\frac{u}{r} + (\lambda + \alpha)\frac{\partial u}{\partial r} - \gamma(T - T_0) \tag{3b}$$

$$\sigma_{zz} = (\lambda + \alpha)\frac{\partial u}{\partial r} + \lambda\frac{u}{r} - \gamma(T - T_0) \tag{3c}$$

$$\sigma_{r\phi} = \sigma_{rz} = \sigma_{\phi z} = 0 \tag{3d}$$

Equations of motion take the form

$$\frac{\partial \sigma_{rr}}{\partial r} + \frac{1}{r}\left(\sigma_{rr} - \sigma_{\phi\phi}\right) = \rho\frac{\partial^2 u}{\partial t^2} \tag{4}$$

Substitution equations (3) into equation (4), we obtain

$$A_{11}\left[\frac{\partial^2 u}{\partial r^2} + \frac{1}{r}\frac{\partial u}{\partial r} - \frac{u}{r^2}\right] - \gamma\frac{\partial T}{\partial r} = \rho\frac{\partial^2 u}{\partial t^2} \tag{5}$$

Where $A_{11} = \lambda + 2\alpha + 4\mu_L - 2\mu_T + \beta$

The relation between the elastic constants and reinforcement parameters is given by [4]

$$2\alpha + \beta + 4(\mu_L - \mu_T) = 0$$

The equation of heat conduction has the form

$$\left[\alpha^* k\frac{\partial}{\partial t} + k^*\right]\nabla^2 T = \rho c_E\frac{\partial^2 T}{\partial t^2} + \gamma T_0\frac{\partial e}{\partial t} \tag{6}$$

where $\alpha^* = 1$ and $\alpha^* = 0$ respectively correspond to heat conduction with and without energy dissipation, k is the thermal conductivity of the medium and c_E is the specific heat at constant strain
The boundary conditions of the problem can be written as:

$$T(r,t) = f(t) \;,\; \sigma_{rr}(r,t) = 0 \qquad at \;\; r = R$$

3. Solution of the Problem in the Laplace Transform Domain

Let us introduce the following non-dimension variables

$$r^* = c_0\eta r\,, u^* = c_0\eta u\,, \sigma^*_{ij} = \frac{\sigma_{ij}}{A_{11}}, \theta = \frac{\gamma(T - T_0)}{A_{11}}, t^* = c_0^2\eta t,\, \eta = \frac{\rho c_E}{k}\;, c_0^2 = \frac{A_{11}}{\rho}$$

The governing equations (3), (5) and (6), in non-dimensional form become (dropping the asterisks for convenience)

$$\sigma_{rr} = \frac{\partial u}{\partial r} + \varepsilon_1 \frac{u}{r} - \theta \tag{7}$$

$$\left[\nabla^2 - \frac{1}{r^2}\right]u - \frac{\partial \theta}{\partial r} = \frac{\partial^2 u}{\partial t^2} \tag{8}$$

$$\left[\alpha^* \frac{\partial}{\partial t} + \varepsilon\right]\nabla^2\theta = \frac{\partial^2 \theta}{\partial t^2} + \varepsilon_2 \frac{\partial^2 e}{\partial t^2} \tag{9}$$

where,

$$\varepsilon = \frac{k^*}{\rho c_E c_0^2} \quad , \quad \varepsilon_1 = \frac{\lambda + \alpha}{A_{11}} \quad , \quad \varepsilon_2 = \frac{\gamma^2 T_0}{A_{11}\rho c_E} \quad ,$$

Introducing the thermoelastic potential function Ψ, defined by $u = \frac{\partial \psi}{\partial r}$

Equations (7)-(9) given by

$$\sigma_{rr} = \frac{\partial^2 \psi}{\partial r^2} + \varepsilon_1 \frac{1}{r}\frac{\partial \psi}{\partial r} - \theta \tag{10}$$

$$\left[\nabla^2 - \frac{\partial^2}{\partial t^2}\right]\psi = \theta \tag{11}$$

$$\left[\alpha^* \frac{\partial}{\partial t} + \varepsilon\right]\nabla^2\theta = \frac{\partial^2 \theta}{\partial t^2} + \varepsilon_2 \frac{\partial^2}{\partial t^2}\nabla^2\psi \tag{12}$$

Applying the Laplace transform with parameter s defined by the relation

$\bar{f}(s)=\int_0^\infty f(t)e^{-st}\,dt$, to both sides of equations (10)-(12), we obtain

$$\bar{\sigma}_{rr} = \frac{\partial^2 \bar{\psi}}{\partial r^2} + \varepsilon_1 \frac{1}{r}\frac{\partial \bar{\psi}}{\partial r} - \bar{\theta} \tag{13}$$

$$\left[\nabla^2 - s^2\right]\bar{\psi} = \bar{\theta} \tag{14}$$

$$\left[\alpha^* s + \varepsilon\right]\nabla^2\bar{\theta} = s^2\bar{\theta} + \varepsilon_2 s^2 \nabla^2 \bar{\psi} \tag{15}$$

Eliminating $\bar{\theta}$ from equations (14) and (15), we get

$$(a\nabla^4 - b\nabla^2 + c)\bar{\psi} = 0 \tag{16}$$

Where $a = \alpha^* s + \varepsilon,\ b = s^2\left(1 + \varepsilon_2 + \alpha^* s + \varepsilon\right),\ c = s^4$

In a similar manner we can show that $\bar{\theta}$ satisfy the equations

$$(a\nabla^4 - b\nabla^2 + c)\bar{\theta} = 0 \tag{17}$$

The solutions of equations (16)-(17), have the forms

$$\bar{\psi} = \sum_{i=1}^{2} A_i K_0(k_i r) \tag{18}$$

$$\bar{\theta} = \sum_{i=1}^{2} A_i\left(k_i^2 - s^2\right)K_0(k_i r) \tag{19}$$

$$\bar{u} = -\sum_{i=1}^{2} A_i\, k_i K_1(k_i r) \tag{20}$$

where $k_1^2\, and\, k_2^2$ are the roots of the characteristic equation.

$$k^6 - ak^4 + bk^2 - c = 0 \tag{21}$$

and $K_0\, and K_1$ is the modified Bessel function of the first kind of order zero.

From equations (18), (19) and (13), we obtain

$$\bar{\sigma}_{rr} = \sum_{i=1}^{2} A_i \left\{ s^2 K_0(k_i r) + [1+\varepsilon_1]\frac{k_i}{r} K_1(k_i r) \right\} \tag{22}$$

Applying the boundary conditions, we obtain the following system of linear equations in the unknown parameters A1 and A2

$$A_1\left(k_1^2 - s^2\right)K_0(k_1 R) + A_2\left(k_2^2 - s^2\right)K_0(k_2 R) = \bar{F}(s) \tag{23}$$

$$A_1\left\{ s^2 K_0(k_1 R) + [1+\varepsilon_1]\frac{k_1}{R} K_1(k_1 R) \right\} + A_2\left\{ s^2 K_0(k_2 R) + [1+\varepsilon_1]\frac{k_2}{R} K_1(k_2 R) \right\} = 0 \tag{24}$$

$$A_1 = \frac{\bar{F}(s)\left\{ s^2 K_0(k_2 R) + [1+\varepsilon_1]\frac{k_2}{a} K_1(k_2 R) \right\}}{\Delta_{11}}$$

$$A_2 = \frac{\bar{F}(s)\left\{ s^2 K_0(k_1 R) + [1+\varepsilon_1]\frac{k_1}{R} K_1(k_1 R) \right\}}{\Delta_{11}}$$

where
$$\Delta_{11} = \left(k_1^2 - s^2\right)K_0(k_1 R)\left\{ s^2 K_0(k_2 R) + [1+\varepsilon_1]\frac{k_2}{a} K_1(k_2 R) \right\} - \left(k_2^2 - s^2\right)K_0(k_2 R)\left\{ s^2 K_0(k_1 R) + [1+\varepsilon_1]\frac{k_1}{R} K_1(k_1 R) \right\}$$

4. Numerical inversion of the Laplace transforms

In order to invert the Laplace transforms, we used the method as in [35]

5. Numerical results

For fixed values of r and t, the parameters A_1 and A_2 are substituted into (19), (20) and (22) to obtain the values of $\bar{\theta}, \bar{u}\, and\, \bar{\sigma}_{rr}$, for different choices for the function $f(t)$ as shown in table 1, where θ_0 and a are constants.

Case	1	2	3
$f(t)$	$\theta_0 H(t)$	e^{-t}	$H(t-a)\sin\left(\frac{\pi t}{a}\right)$

TABLE 1

The solution in the space–time domain is obtained numerically on employing the numerical inversion of the Laplace transform proposed by FORTRAN software is used for the computational purpose. For the computation, we have used carbon fiber–epoxy resin composite. The values of the physical data are given by [27, 28]

$$\lambda = 5.65\text{x}10^9 \text{ N/m}^2, \mu_T = 2.46\text{x}10^9 \text{ N/m}^2, \mu_L = 5.66\text{x}10^9 \text{ N/m}^2, \alpha = -1.28\text{x}10^9 \text{ N/m}^2,$$
$$\beta = -10.24\text{x}10^9 \text{ N/m}^2, \rho = 1660\,\text{kg/m}^3, \alpha_t = 1.78\text{x}10^{-5} \text{ K}^{-1}, C_E = 797\,\text{J.kg}^{-1}.\text{K}^{-1},$$
$$T_0 = 293\text{K}, \text{ K} = 386\,\text{W.mK}^{-1}, k^* = 130\,\text{W.mK}^{-1},$$

We shall now compare, graphically, our results, where the theory of thermoelasticity with and without energy dissipation (GN theory II, III) and the generalized theory of thermoelasticity with one relaxation time due to Lord-Shulman (LS theory). The calculations are carried out for a time of $t = 0.3$ The Solid line represents the solution for thermoelasticity with energy dissipation (G-N II), dashed lines represent the solution for thermoelasticity without energy dissipation (G-N III) and Dotted lines represent the solution for generalized theory of thermoelasticity with one relaxation time (L-S).

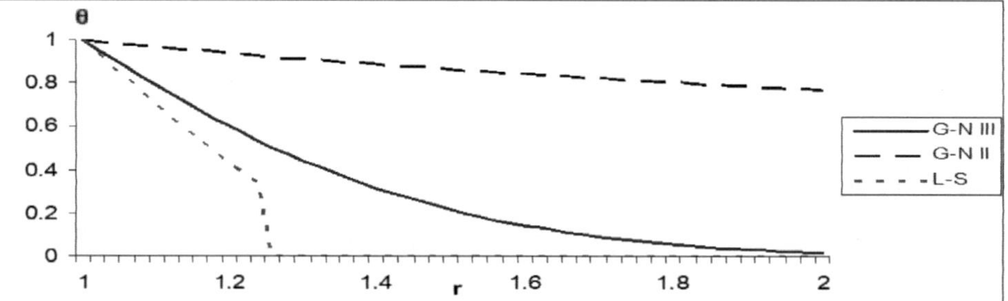

Fig. 2 Temperature Distribution

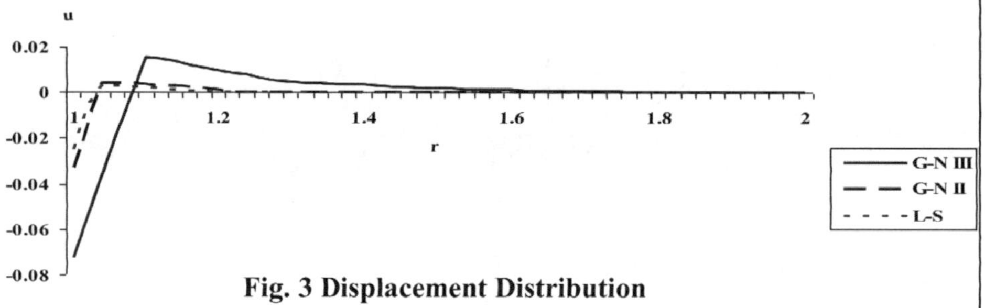

Fig. 3 Displacement Distribution

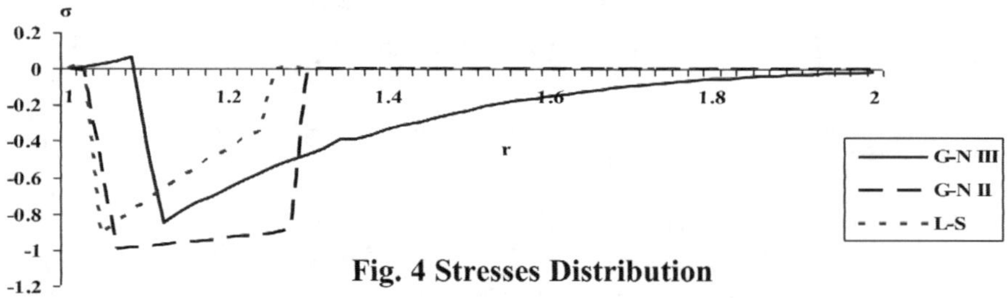

Fig. 4 Stresses Distribution

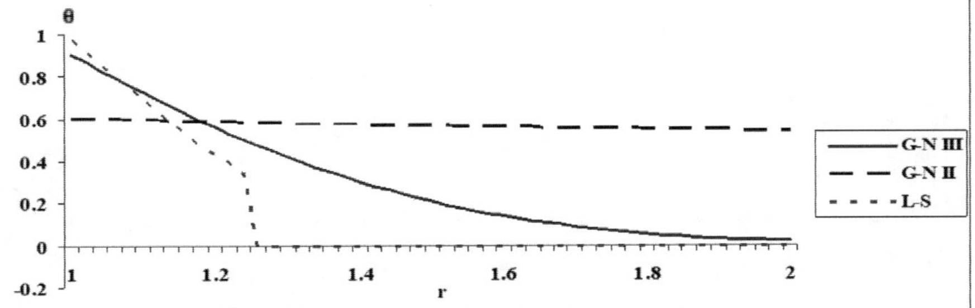

Fig. 5 Temperature Distribution

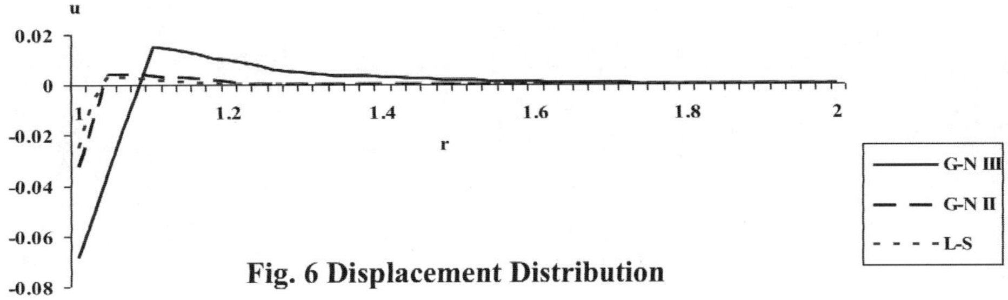

Fig. 6 Displacement Distribution

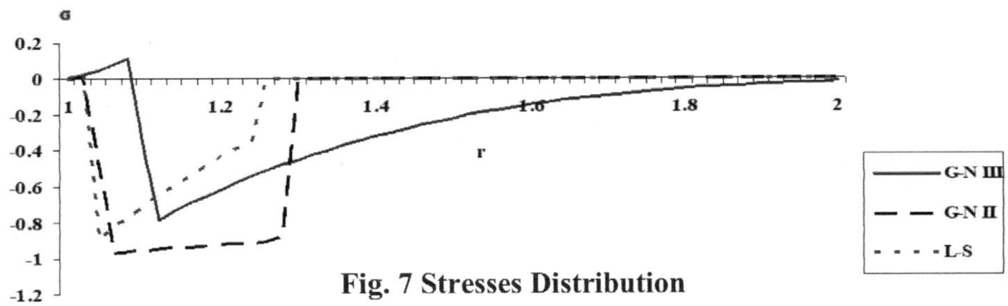

Fig. 7 Stresses Distribution

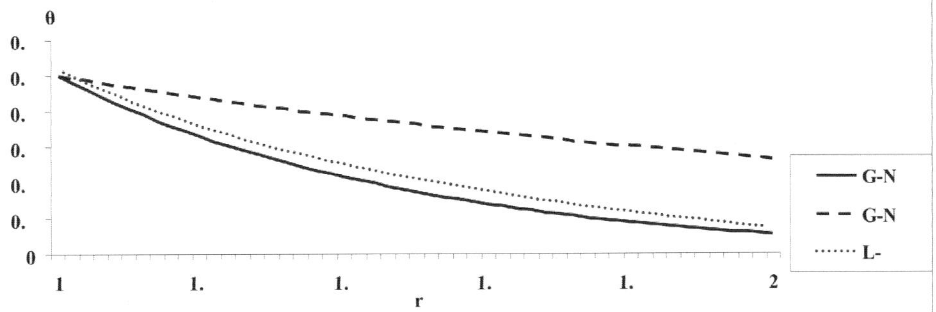

Fig. 8 Temperature Distribution

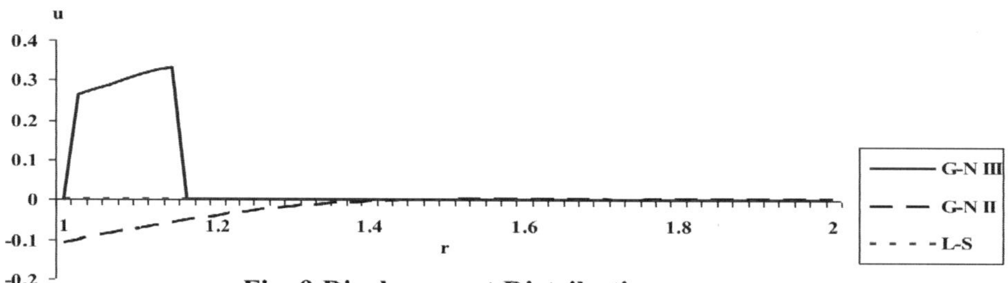

Fig. 9 Displacement Distribution

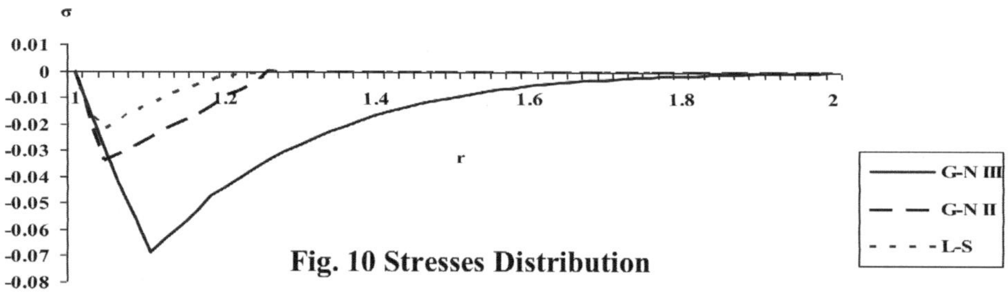

Fig. 10 Stresses Distribution

Figures (2-4) are representing the first case, while the figures (5-7) are representing the second case, and the figures (8-10) are representing the third case. For the three cases, the temperature distributions are shown in Figures 2, 5, and 8, the displacement distributions are shown in Figures 3, 6, and 9, while the stress distributions are shown in Figures 4, 7, and 10. In figures 2, 5 and 8 shows the variation of temperature with r and it indicates that temperature field has maximum value at the boundary $r = 1$ and then decreases to zero. For example, in Fig. 2 at $r = 1.225$ in the case of (L-S), and $r = 2$ in the case of (G-N III) while in the case of (G-N II) the energy loss is very little. Figures 3, 6 and 9 shows the variation of displacement with r, and have the same behaviour in the first and second cases, while in the third case the values of displacement are very small in (L-S) with respect to (G-N). Figures 4, 7 and 10 shows the variation of stress with r and it indicates the stress field has zero value at the boundary $r = 1$, it is clear that there is a discontinuity in the stress in case one and two but in third case the discontinuity is not shown because it is very small.

6. Concluding remarks

The present study aims to studying the wave propagation for a problem of cylindrical cavity, the material is assumed to be fiber-reinforced with increased stiffness, strength and load bearing capacity. Using generalized thermoelasticity theory, Green and Nagdhi, with and without energy dissipation, we can obtain the following conclusions based on the above analysis.

The discontinuity is clearly shown in G-N III and LS theory models compared to G-N II model. The reinforcement has a great effect on the distribution of field quantities. The results presented in this paper should beneficial to researchers in material science, designers of new materials.

References

[1] M. S. CRAIG, V. G. HART, "The Stress Boundary-Value Problem for Finite Plane Deformations of a Fiber-Reinforced Material", The Quarterly Journal of Mechanics and Applied Mathematics, Vol. 32, No. 4, pp. 473–498.

[2] B. Singh, S. J. Singh, 2004, "Reflection of plane waves at the free surface of a fibre-reinforced elastic half-space", Sadhana Vol. **29,** pp. 249–257.

[3] F. S. Bayones, N. S. Hussein, 2015, "Fiber-reinforced generalized thermoelastic medium subjected to gravity field", Journal of Modern Physics, Vol. 6, No. 8, pp. 1113–1134.

[4] S. Mondal, A. Sur, M. Kanoria, 2019, "Magneto-thermoelastic interaction in a reinforced medium with cylindrical cavity in the context of Caputo–Fabrizio heat transport law" Acta Mech., Vol. 230, pp. 4367–4384.

[5] I. A. Abbas, M. I. A. Othman, 2012, "Generalized thermoelastic interaction in a fiber-reinforced anisotropic half-space under hydrostatic initial stress", J. Vib. Control, Vol. 18, No 2, pp. 175-182.

[6] A. E. Abouelregal, A. M. Zenkour, 2014, "On the generalized thermoelasticity problem for an infinite fiber-reinforced thick plate under initial stress", Adv. Appl. Math. Mech., Vol. 6, No. 6, pp. 783–796.

[7] A. Sur, M. Kanoria, 2015, "Fibre-reinforced magneto-thermoelastic rotating medium with fractional heat conduction", Procedia Eng., Vol. 127, No. 3, pp. 605–612.

[8] A. Sur, P. Pal, M. Kanoria, 2018, "Modeling of memory-dependent derivative in a fiber-reinforced plate under gravitational effect" J. Therm. Stress, Vol. 41, No. 8, pp. 973–992.

[9] B. Singh, 2006, "Wave propagation in thermally conducting linear fiber-reinforced composite materials", Archive of Applied Mechanics, Vol. 75, No. 5, pp. 513–520.

[10] Heung Soo Kim, Jung Woo Sohn, Seung-Bok Choi, 2011, "Vibration Control of a Cylindrical Shell Structure Using Macro Fiber Composite Actuators", Mechanics Based Design of Structures and Machines, Vol. 39, No. 4, pp. 491-506.

[11] Ibrahim A. Abbas, 2015, "A Dual Phase Lag Model on Thermoelastic Interaction in an Infinite Fiber-Reinforced Anisotropic Medium with a Circular Hole", Mechanics Based Design of Structures and Machines, Vol. 43, No. 4, pp. 501-513.

[12] A. J. Belfield, T. G. Rogers, A. J. M Spencer, 1983, "Stress in elastic plates reinforced by fibers lying in concentric circles", Journal of the Mechanics and Physics of Solids, Vol. 31, No. 1, pp. 25-54.

[13] R. Kumar, R. R. Gupta, 2010, "Study of wave motion in an anisotropic fiber-reinforced thermoelastic solid", Journal of Solid Mechanics, Vol. 2, No. 1, pp. 91–100.

[14] M. I. A. Othman, S. M. Said, N. Sarker, 2013, "Effect of hydrostatic initial stress on a fiber-reinforced thermoelastic medium with fractional derivative heat transfer", Multidiscipline Modeling in Materials and Structures, Vol. 9, No. 3, pp. 410–426.

[15] J. P. Boehler, J. Raclin, 1985, "Failure Criteria for Glass-Fiber Reinforced Composites under Confining Pressure", Mechanics Based Design of Structures and Machines, Vol. 13, No. 3, pp. 371-393.

[16] H. Lord, Y. Shulman, 1967, "A generalized dynamical theory of thermoelasticity", J Mech Phys Solid, Vol. 15, pp. 299–309.

[17] Hany Sherief, Nasser M El-Maghraby, 2003, "An Internal Penny-Shaped Crack in an Infinite Thermoelastic Solid", J Thermal Stresses, Vol. 26, No. 2, pp. 333-352.

[18] Hany H Sherief, Nasser M El-Maghraby, 2005, "A Mode-I Crack Problem for an Infinite Space in Generalized Thermoelasticity", J. Thermal Stresses, Vol. 28, No. 3, pp. 465-484.

[19] Hany Sherief, Farid Hamza, 1996, "Generalized Two-Dimensional Thermoelastic Problems in Spherical Regions under Axisymmetric Distributions", J. Thermal Stresses, Vol. 19, No. 1, pp. 55-76.

Journal of Physics: Conference Series **2304** (2022) 012005 doi:10.1088/1742-6596/2304/1/012005

[20] Hany Sherief, Farid Hamza, 1994, "Generalized Thermoelastic Problem of a Thick Plate Under Axisymmetric Temperature Distribution", J. Thermal Stresses, Vol. 17, No. 4, pp. 435-452.

[21] Hany Sherief, Magdy A Ezzat, 1994, "Solution of the Generalized Problem of Thermoelasticity in the form of Series of Functions", J. Thermal Stresses, Vol. 17, NO. 2, pp. 75-95.

[22] Hany Sherief, Ranjit Dhaliwal, 1981, "A Generalized One-Dimensional Thermal Shock Problem for Small Times", J. Thermal Stresses, Vol. 4, No. 4, pp. 407-420.

[23] Hany H Sherief, Farid A Hamza, Amany M El-Sayed, 2005, "Theory of Generalized Micropolar Thermoelasticity and an Axisymmetric Half-Space Problem", J. Thermal Stresses, Vol. 28, No. 2, pp. 409-437.

[24] Hany H Sherief, Mohammed N Allam, Mohammed A El-Hagary, 2011, "Generalized Theory of Thermoviscoelasticity and a Half-Space Problem", Int. J. Thermophys., Vol. 32, No. 5, pp. 1271–1295.

[25] A. Montanaro, 2011, "On piezo thermoelastic plates subject to prescribed boundary temperature", Meccanica, Vol. 46, No. 6, pp. 383-398.

[26] S. Chirita, 2012, "On the final boundary value problems in linear thermoelasticity", Meccanica, Vol. 47, No. 3, pp. 2005-2011.

[27] A. Nayfeh, S. Nemat-Nasser, 1972, "Electromagneto-thermoelastic plane waves in solids with thermal relaxation", J. Appl. Mech. Series E, Vol. 39, No. 3, pp. 108-113.

[28] Hany Sherief, S E Khader, 2013, "Propagation of discontinuities in electromagneto generalized thermoelasticity in cylindrical regions", Meccanica, Vol. 48, No. 5, pp. 2511–2523.

[29] R. Kumar, S. Devi, 2019, "Deformation of modified couple stress thermoelastic diffusion in a thick circular plate due to heat sources", CMST, Vol. 25, No. 4, pp. 167–176.

[30] R. Kumar, S. Devi, S. M. Abo-Dahab, 2021, "Propagation of Rayleigh waves in modified couple stress generalized thermoelastic with three-phase-lag model", Waves in random and complex media, Vol. 31, No. 2, 359-371.

[31] H. H. Sherief, H. M. Yosef, 2004, "Short Time Solution for a Problem in Magneto-thermoelasticity with Thermal Relaxation", J. Thermal Stresses, Vol. 27, No. 3, pp. 537-559.

[32] J. N. Sharma, D. Chand, 1988, "Transient generalized magneto thermoelastic waves in a half space", Int. J. Eng. Sci., Vol. 26, No. 4, pp. 951–958.

[33] A. M. Abd El-Latief, S. E. Khader, 2015, "Fractional model of thermoelasticity for a half-space overlaid by a thick layer", Z. angew. Math. Mech., Vol. 95, No. 4, pp. 511-518.

[34] M. I. A. Othman, 2011, "State space approach to the generalized thermoelastic problem with temperature-dependent elastic moduli and internal heat sources", Journal of Applied Mechanics and Technical Physics, Vol. 52, No. 5, pp. 644–656.

[35] G. Honig, U. Hirdes, 1984, "A Method for The Numerical Inversion of the Laplace Transform", J. Comp. Appl. Math., Vol. 10, No. 2, pp. 113-132.

A Decoupling and Coupling Approach for the Bi-harmonic Equation

Kamal Hassan[1], Mahmoud Farid [2]

[1,2] Department of basic science, the British University in Egypt, Cairo, Egypt
[1]Email:Kamal.Hassan@bue.edu.eg

[2]Email:Mahmoud.farid @bue.edu.eg

Abstract. This paper is devoted to the numerical solution of the bi-harmonic problem: $\boldsymbol{\Delta^2 w = p}$. The operator is split into two harmonic ones. The finite difference scheme of the harmonic operator is used to get discretized operators, coupling those leads to a global system $\boldsymbol{A\vec{w} = \vec{p}}$, this system is to be solved with any well-posed boundary conditions. In this way, boundary conditions need not be specified with the decoupled equations, but rather with the global system. Numerical test problems are provided. The extension of this technique to poly harmonic equations is straightforward.

1. Introduction

Let Ω be a smooth bounded domain in R^2. Denote by Γ the boundary of Ω and by n the unit outward direction normal to Γ. The aim of this study is to exploit the methods of discretization of Laplace operator for the discretization of the biharmonic operator, paving the way for polyharmonic ones. So, this paper will handle the approximation of the biharmonic equation:

$$\Delta^2 w = p \quad \text{In } \Omega \tag{1}$$

Subject to well-posed boundary conditions. The most famous ones are:

$$w = \frac{\partial w}{\partial n} = 0 \qquad \text{in } \Gamma \tag{2a}$$

$$w = \frac{\partial^2 w}{\partial n^2} = 0 \qquad \text{in } \Gamma \tag{2b}$$

$$w = \frac{\partial^3 w}{\partial n^3} = 0 \qquad \text{in } \Gamma \tag{2c}$$

It is worth noting that the system of equations (1) and (2a) determine the transverse deflection w of a clamped thin plate loaded by a resultant transverse force p while the system (1) and (2b) represents the simply supported plate.

There are different methods of solving these equations [1]. In this paper, we focus on the finite-difference method [see [2] for different schemes].and more precisely on the indirect approach in which the bi-harmonic equation is split into two coupled Poisson equations. Thus we can exploit the simplicity and accuracy of the solution of the Poisson equation [3].

There are many different high order accurate finite difference discretization schemes for the harmonic operator. In order to check the validity of the proposed tool, the traditional five-point second-order finite difference approximation to the Laplacian is applied.

Equation (1) is replaced by the two Poisson equations:

$$\begin{cases} \Delta w = v \\ \Delta v = p \end{cases} \tag{3}$$

Then, these equations are to be solved. This goes well with the boundary conditions (2b), there are boundary conditions for w and v (imposing w=0 with $\frac{\partial^2 w}{\partial n^2} = 0$ implies that Δw =0, so $v = 0$). The two Poisson equations are solved in a straightforward way. But with the other boundary conditions, (2a, 2c) we face the difficulty of the non-existence of boundary conditions for v.

Therefore the function v needs to be approximated from the discrete form of $\Delta w = v$ [4-7]. Moreover, there may also be a need for matrix decomposition and Fast Fourier Transforms with this case [8].It is to be noticed that knowing v is of practical importance because it has a physical meaning such as bending moment in plate deflection.

In this work, we propose to apply the decoupling and use the finite-difference to get the matrix representation of (3)

$$\begin{cases} \mathrm{A}\vec{w} = \vec{v} \\ \mathrm{A}\vec{v} = \vec{p} \end{cases} \tag{4}$$

Combining the two expressions of (4) leads to one global matrix equation:

$$\mathrm{A}(\mathrm{A}\vec{w}) = \mathrm{A}^2\vec{w} = K\vec{w} = \vec{p} \tag{5}$$

Where $K = \mathrm{A}^2$ Equation (5) is to be solved with any of the boundary conditions (2a or 2b or 2c).

It is worth noting that once the values of w are determined, those of v can be obtained through the relation $\boldsymbol{A}\vec{\boldsymbol{w}} = \vec{\boldsymbol{v}}$. The above approach is not valid for finite difference schemes which has the matrix form $\boldsymbol{A}\vec{\boldsymbol{x}} = \boldsymbol{B}\vec{\boldsymbol{y}}$ [9]. Research is going on to handle these situations.

2. Method of Solution

To explain and verify the validity of the proposed approach, the following steps are followed:

Step 1: The Finite Difference Scheme

We consider our region of interest, a rectangular domain $\Omega= [0.1] \times [0.1]$.This domain is divided into N points with increments in the x and y direction as Δx and Δy where, $\Delta x = \Delta y = h$. Each nodal point is designated by a numbering i and j where i and j are x and y increments respectively as shown in Fig.1.

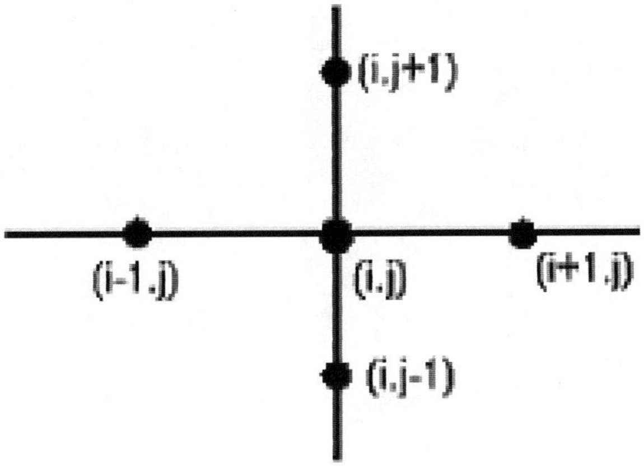

Fig. 1. Finite Difference Scheme

The central difference approximation for second order derivatives is given by:

$$\left(\frac{\partial^2 w}{\partial x^2}\right) = \frac{1}{h^2}\left(w_{i-1.j} - 2w_{i,j} + w_{i+1.j}\right) + O(h^2) \tag{6a}$$

$$\left(\frac{\partial^2 w}{\partial y^2}\right) = \frac{1}{h^2}\left(w_{i.j-1} - 2w_{i.j} + w_{i.j+1}\right) + O(h^2) \tag{6b}$$

Adding the above equations lead to:

$$\begin{aligned}\Delta w = \left(\frac{\partial^2 w}{\partial x^2} + \frac{\partial^2 w}{\partial y^2}\right) &= \frac{1}{h^2}\left(w_{i-1.j} - 2w_{i.j} + w_{i+1.j}\right) + \frac{1}{h^2}\left(w_{i.j-1} - 2w_{i.j} + w_{i.j+1}\right) \\ &= \frac{1}{h^2}\left(w_{i-1.j} + w_{i.j-1} - 4w_{i.j} + w_{i+1.j} + w_{i.j+1}\right)\end{aligned} \tag{7}$$

As an illustrative example the case of N=3 is given in detail with the numbering shown in Fig. 2

Journal of Physics: Conference Series **2304** (2022) 012006 doi:10.1088/1742-6596/2304/1/012006

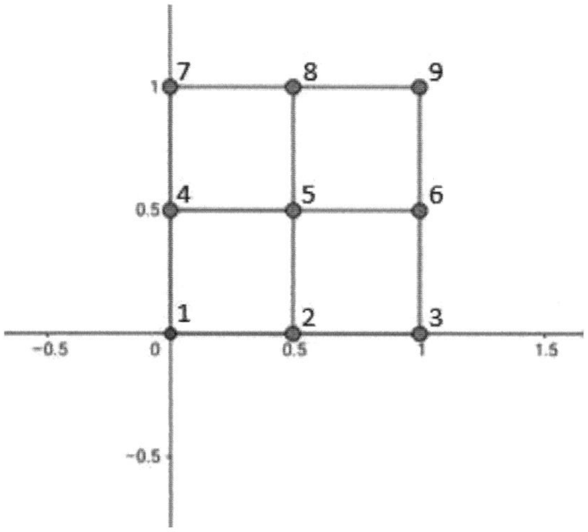

Fig.2. Finite Difference nodal numbering

Step 2: The matrix A

$$\begin{bmatrix} -4 & 1 & 0 & 1 & 0 & 0 & 0 & 0 & 0 \\ 1 & -4 & 1 & 0 & 1 & 0 & 0 & 0 & 0 \\ 0 & 1 & -4 & 0 & 0 & 1 & 0 & 0 & 0 \\ 1 & 0 & 0 & -4 & 1 & 0 & 1 & 0 & 0 \\ 0 & 1 & 0 & 1 & -4 & 1 & 0 & 1 & 0 \\ 0 & 0 & 1 & 0 & 1 & -4 & 0 & 0 & 1 \\ 0 & 0 & 0 & 1 & 0 & 0 & -4 & 1 & 0 \\ 0 & 0 & 0 & 0 & 1 & 0 & 1 & -4 & 1 \\ 0 & 0 & 0 & 0 & 0 & 1 & 0 & 1 & -4 \end{bmatrix}$$

Step 3: The Matrix A^2

$$\begin{bmatrix} -18 & -8 & 1 & -8 & 2 & 0 & 1 & 0 & 0 \\ -8 & 19 & -8 & 2 & -8 & 0 & 0 & 1 & 0 \\ 1 & -8 & 18 & 0 & 2 & 1 & 0 & 0 & 1 \\ -8 & 2 & 0 & 19 & -8 & 0 & -8 & 2 & 0 \\ 2 & -8 & 2 & -8 & 20 & 1 & 2 & -8 & 2 \\ 0 & 2 & -8 & 1 & -8 & -4 & 0 & 2 & -8 \\ 1 & 0 & 0 & -8 & 2 & 0 & 18 & -8 & 1 \\ 0 & 1 & 0 & 2 & -8 & 0 & -8 & 19 & -8 \\ 0 & 0 & 1 & 0 & 2 & -8 & 1 & -8 & 18 \end{bmatrix}$$

Step 4: The boundary conditions for the case (2a):

$$w = \frac{\partial w}{\partial n} = 0 \qquad \text{in } \Gamma$$

We will use the central finite difference scheme second order. Thus we keep the same order of accuracy as that of Laplacian operator

$$\frac{\partial w_{i.j}}{\partial n} = \frac{w_{i+1.j} - w_{i-1.j}}{2h} + O(h^2) = 0 \tag{8a}$$

$$\left(\frac{\partial^2 w}{\partial y^2}\right) = \frac{1}{h^2}\left(w_{i.j-1} - 2w_{i.j} + w_{i.j+1}\right) + O(h^2) \tag{8b}$$

Then $w_{i+1.j} = w_{i-1.j}$ and $w_{j+1.j} = w_{j-1.j}$ on the boundary, therefore the matrix

A^2 Will now take the form:

$$\begin{bmatrix} 18 & -16 & 1 & -16 & 2 & 0 & 1 & 0 & 0 \\ -8 & 19 & -8 & 2 & -16 & 2 & 0 & 1 & 0 \\ 1 & -16 & 18 & 0 & 2 & -16 & 0 & 0 & 1 \\ -8 & 2 & 0 & 19 & -16 & 1 & -8 & 2 & 0 \\ 2 & -8 & 2 & -8 & 20 & -8 & 2 & -8 & 2 \\ 0 & 2 & -8 & 1 & -16 & 19 & 0 & 2 & -8 \\ 1 & 0 & 0 & -16 & 2 & 0 & 18 & -16 & 1 \\ 0 & 1 & 0 & 2 & -16 & 2 & -8 & 19 & -8 \\ 0 & 0 & 1 & 0 & 2 & -16 & 1 & -16 & 18 \end{bmatrix}$$

Step 5: Solve the system $A^2\vec{w} = \vec{p}$

3. Numerical Results

The square $\Omega = [0.1] \times [0.1]$ is divided into (N-1) intervals in each direction. ***MATLAB*** code is used to get the numerical results

3.1. Test Problem 1:

The solution is analytically known [10]: $W = x^3(1-x)^3 y^3(1-y)^3$

The Right Hand Side is chosen so as to give a pre- assigned solution.

$$\begin{gathered} p = -72(1-x)^2y^3(1-y)^3 + 216x(1-x)y^3(1-y)^3 \\ -72x^2y^3(1-y)^3 + 72x(1-x)^3y(1-y)^3 \\ -216x(1-x)^3y^2(1-y)^2 + 72x(1-x)^3y^3(1-y) \\ -216\,x^2(1-x)^2y(1-y)^3 + 648x^2(1-x)^2y^2(1-y)^2 \\ -216x^2(1-x)^2y^3(1-y) + 72x^3(1-x)y(1-y)^3 \\ -216x^3(1-x)y^2(1-y)^2 + 72x^3(1-x)y^3(1-y) \\ -72x^3(1-x)^3(1-y)^2 + 216x^3(1-x)^3y(1-y) - 72x^3(1-x)^3y^2 \end{gathered} \tag{9}$$

The Exact solution at the centre of the domain is **0.000244140625**

Table 1. Finite Difference solution of Test Problem 1 at the middle of the domain

N	Approximate solution	N	Approximate solution
5	0.000474078745781621	115	0.000244230995398045
15	0.000253157361519303	125	0.000244216577973587
25	0.000246715677559747	135	0.000244205352021139
35	0.000245323758183147	145	0.000244196441177007
45	0.000244814673263896	155	0.000244189250329623
55	0.000244574622084671	165	0.000244183363914567
65	0.000244442985034806	175	0.000244178484638850
75	0.000244363183412735	185	0.000244174395255570
85	0.000244311217523361	195	0.000244170934121341
95	0.000244275513462870	205	0.000244167978863782
105	0.000244249937632485		

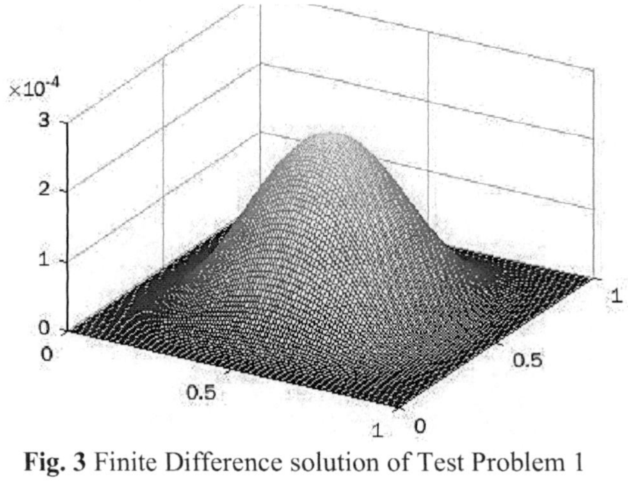

Fig. 3 Finite Difference solution of Test Problem 1

3.2. Test Problem 2:

P=1, Numerical solution at the centre of the domain is $W =$ 0.00126532 [11]

Table 2. Finite Difference solution of Test Problem 2 at the middle of the domain

N	Approximate solution	N	Approximate solution
5	0.00243522547468354	405	0.00127785081377727
45	0.00138057723004942	445	0.00127672152803582
85	0.00132564878787286	485	0.00127577894371635
125	0.00130617310295876	525	0.00127498029924163
165	0.00129620248175754	565	0.00127429496294969
205	0.00129014370760939	605	0.00127370040998693
245	0.00128607223757230	645	0.00127317973833385
285	0.00128314808613431	685	0.00127271997321548
325	0.00128094618980118	705	0.00127250967996316
365	0.00127922837264005		

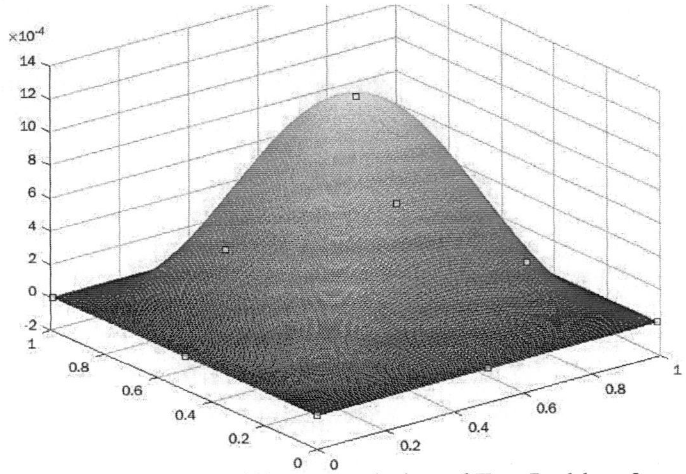

Fig. 4. Finite Difference solution of Test Problem 2

It is clear that the least accurate finite difference scheme converges to the solution.

4. Conclusion and future work

It is shown that the decoupling of the continuous bi-harmonic operator enables the use of the finite difference schemes of the harmonic operator to solve the bi-harmonic equation. Coupling of the discretized operators can handle any well-posed boundary conditions with the bi-harmonic operator. An extension of the present study to poly-harmonic equations is straightforward. Currently, we are working on solving the eighth-order elliptic partial differential equation with different boundary conditions.

The success of applying the proposed approach to finite element techniques implies the formation of the matrices of equations (4) and (5) using usual C^0 finite elements. Thus we can avoid the necessity of C^1 continuity of the transverse deflection which is required for displacement type of finite elements [12].

5. Acknowledgment

The authors thank professor Talat Refai of the British University in Egypt for his useful comments.

6. References

[1] Bjorstad P.F. 1980, Numerical solution of the bi-harmonic equation, Report, No.STAN-CS, 80-834, Department of computer science, Stanford University.

[2] Jovanovic´B.S., Süli E. 2014, Analysis of Finite Difference Schemes for Linear Partial Differential Equations with Generalized Solutions. Springer-Verlag London.

[3] Mohanty.R.K, Singh S., 2006, A new fourth order discretization for singularly perturbed two dimensional non-linear elliptic boundary value problems, Applied Mathematics and Computation vol 175 P. 1400–1414.

[4] Dehghan M., Mohebbi 2008, A. Solution of the two dimensional second bi-harmonic equation with high-order accuracy. Kybernetes , 37 (8); 1165-1179.

[5] Smith J. On the approximate solution of the first boundary value problem for $\nabla^4 u = f$, SIAM J.Numer Anal. 1973; 10:967-982.

[6] Ehrlich L.W, 1971, solving the bi-harmonic equation as coupled finite difference equations. SIAM J. Numer. Anal. (8); 278–287.

[7] Gupta M.M., Manohar R.P., 1985, Stephenson, J. W. High-order difference schemes for two-dimensional elliptic equations. Numerical Methods for Partial Differential Equations. (1); 71–80.

[8] Bialecki B, Fairweather G, Karageorghis A, Maack J. 2020, A quadratic spline collocation method for the Dirichlet biharmonic problem, Numerical Algorithms; 83:165–199.

[9] Wang T., Liu T., R.K, Singh S., 2019, A Consistent Fourth-Order Compact Finite Difference Scheme for Solving Vorticity-Stream Function Form of Incompressible Navier-Stokes Equations, Numer. Math. Theor. Meth. Appl. 12(1); 312-330.

[10] Chinosi C., Lovadina C., 1995, Numerical analysis of some mixed finite element methods for Reissner–Mindlin plates, Comput.Mech. 16, 36–44.

[11] Timoshenko, S.P. and S. Woinowsky-Krieger, 1959, Theory of plates and shells. Mc Graw-hill.

[12] Bathe KJ. 1996, Finite element procedures. Englewood Cliffs, NJ: Prentice-Hall.

Multi Chaotic System to Generate Novel S-Box for Image Encryption

Hany Nasry[1], Azhaar A. Abdallah[2], Alaa K. Farhan[3], Hossam E. Ahmed[4], Wageda I.El Sobky[5,6]

[1] *Mathematical Departement ,Military Technecial Collage ,Cairo,Egypt[1]*
Department of Applied Sciences, University of Technology, Baghdad, Iraq[2,3]
Department of Communication Engineering, Benha Faculty of Engineering, Benha University, Egypt[4]
Department of *Basic* Engineering science, Behna Faculty of Engineering, Benha University, Egypt[5]
Department of Basic Science, Canidian International College (CIC), Egypt6.
Corresponding author[1]: hanynasry@mtc.edu.eg
Corresponding author[3]: 110030@uotechnology.edu.iq
Corresponding author[5]: wageda.alsobky@bhit.bu.edu.eg

Abstract. A novel method on the basis of multi chaos theory is suggested in the presented study. Also, the study used two different dimensions to generate S-Box to get a strong cipher that is difficult to break. The suggested image cryptosystem includes an identical (decryption and encryption) process, which involves a single keystream generator, shifting process (based on 3D Lorenz map) related diffusion operations, and generate S-Box (based on 2D Henon map) that related confusion operation. The comparative analysis and the simulate test show that the suggested image cryptosystem has a few properties, like high-sensitivity, fast encryption/decryption, large keyspace, excellent statistical properties related to the ciphertext, and so on. The suggested cryptosystem is considered as an alternative for practical secure communications.

Keywords: Chaos System, Image Encryption, S-Box, Shifting

1. INTRODUCTION

The research on image encryption is an important point in current encryption algorithm [1,2] after AES and DES became standards in terms of encrypting the text data. Also, the experts of cryptography attempt to identify the optimum algorithm of image encryption for serving as an algorithm for image encryption standard. In comparison to the text data, the properties of digital images, including strong correlation, large data amounts, big data redundancy, make it need a large quantity of pseudo-random numbers as a keystream, which is, simulating the researches on pseudo-random numbers. In addition, the chaotic systems are created via deterministic equations with vital benefits to create pseudo-random numbers since they were qualified with maximum sensitivity to the initial values as well as parameters, ergodicity, dense form, and so on. Currently, the chaotic systems were majorly utilized in the systems of image encryption as key stream generators [3,4], and extended with multiple images [5], in A.I also using the chaotic system to generated ranomen population and using in another application [6] might be categorized into 2 categories: 1D chaotic maps [7,8] and multi-dimensional (MD) chaotic maps [7,6]. Normally, 1D chaotic maps consist of a single variable [42-46] and a number of parameters. For instance, the Logistic, Sine, and Tent maps and can combine multiple chaotic maps as chaotic hybrid maps [9,10].

In this paper, to transfer the data into a perplexing type, such as block ciphers are using two major permutation and substitution operations. A substitution procedure uses a substitution table referred to as the substitution box (S-box) for replacing byte/block with another one [11,12]. On the other hand, in some linear methods, a permutation method shifts the input bits or bytes [13].

For producing strong S-boxes, researchers and academics have explored and examined different concepts [14]. The intensity was evaluated using some usual parameters, like non-linearity, lack of fixed points, differential and linear probabilities, the strict criterion of avalanche (SAC), the criterion of bit independence (BIC), and so on. [15]. In addition, section two is providing an overview of the chaos theory (2-D and 3-D). In Section 3, a novel approach for image encryption based on generated S-Box with shifting, also a performance evaluation as well as a comparison of encryption was performed. The research findings are provided in Section.

2. Chaotic Map

Chaos theory depends on (initial & condition) parameter sensitivity. This means that any small change gives different results.

A- 2D - The Hénon Map A discrete time dynamic system was often called Hénon-Pomeau attractor/map. It can be specified as a majorly examined example of chaotic behavior displaying dynamical systems. Furthermore, the Henon system is taking a point (x_n, y_n) in the plane and after that mapping it into a new point [1][16]. Introduced 2D-Henon system., as described in Eq. (1).

$$x_{i+1} = 1 - ax_i^2 + y_i \tag{1}$$

$$y_{i+1} = b\, x_i$$

The map is based on 2 parameters, a and b, with values of a = 1.4 and b = 0.3 for the classical Hénon system [17].

B- 3D - Lorenz System the Lorenz system [18,19], which Edward Lorenz researched for the first time in 1960, is a dynamic system defined by the nonlinear system of ordinary equations:

$$X_{n+1} = \alpha(Y_n - Z_n) \tag{2}$$

$$Y_{n+1} = RX_n + X_nZ_n - Y_n$$

$$Z_{n+1} = X_nY_n - BZ_n$$

These variables (α, r, b) are referred to as control parameters, while (x, y, z) is referred to as status variables [20]. Equation (2) defines the control parameters, and the initial values x_0, y_0, z_0 are referred to as state variables, and they are 10, 8/3, and 28, respectively.

3. THE PROCESS OF ENCRYPTION & DECRYPTION

Because of the strong correlations amongst neighbour pixels of the plain image, this study suggests shifting the pixel positions related to plain image for solving such problem as well as breaking the pixels' correlations. Without a generality loss, the plain image dimensions are going to be N × N. The proposed method focuses on the principles (diffusion and confusion) for breaking the pixels' correlations. Design a new S-Box that gives a better way of Shifting to more confusion and diffusion.

4. S-Box Generator

There are several methods of generating S-Boxes [32-41].This work suggested designing a new S-Box depending on 2D-Henon map; this operation provides greater protection and complexity to generate new large $(\mathbf{16 \times 16})$. Initially, using initial value $\mathbf{X_0}$ to chaotic map and creating numbers, the range $(\mathbf{0} - \mathbf{255})$, S-Box production mechanism by generating Henon system values, all S-Box values must be unique. If the value is greater than the appropriate area, then the rest of the section has been taken to that value, producing S-Box inverse at the same time depending on the result of S-Box, since the "responsive dependency on 5 initial states" with chaos theory changes the construction of S-Box and the result of dynamic S-Box inverse with every slight change in initial value.

5. Encryption Schema Steps

The details of image encryption are indicated in the following way **(as shown in fig. 1)**:

Step 1 Shifting columns based on the value of X_n that's generated from 3D-Lornez System then XORed the result matrix with the value of X_n

Step 2 Shift rows based on the value of Yi that's generated from 3D-Lornez System and the result matrix XORed with the value of Yi.

Step 3 To increase diffusion principle in pixel, replace rows with columns then XORed the result matrix with the value of Zi that generated from 3D-Lornez System.

Step 4 For more confusion, generate a new S-Box (16*16) depend on the 2D-Henon map in unique values and non-linearity, then inserting S-Box to the above matrix.

Step 5 Finally, we get cipher and ambiguous images.

6. Decryption Schema Steps

The details of image decryption are indicated in the following way:

Step 1 Substitute the image matrix based on inverse S-Box generated from 2D-Henon map.

Step 2 XOR result matrix with Zi that generated from 3D-Lorenze map then replace rows & columns.

Step 3 XOR result matrix with Yi then shifts columns based on Yi (Yi generated from 3D-Lorenz map).

Step 4 XOR result matrix with X_i then shifts columns based on X_i (X_i generated from 3D-Lorenz map).

Step 5 Finally, extract the plain image with a slight difference in resolution

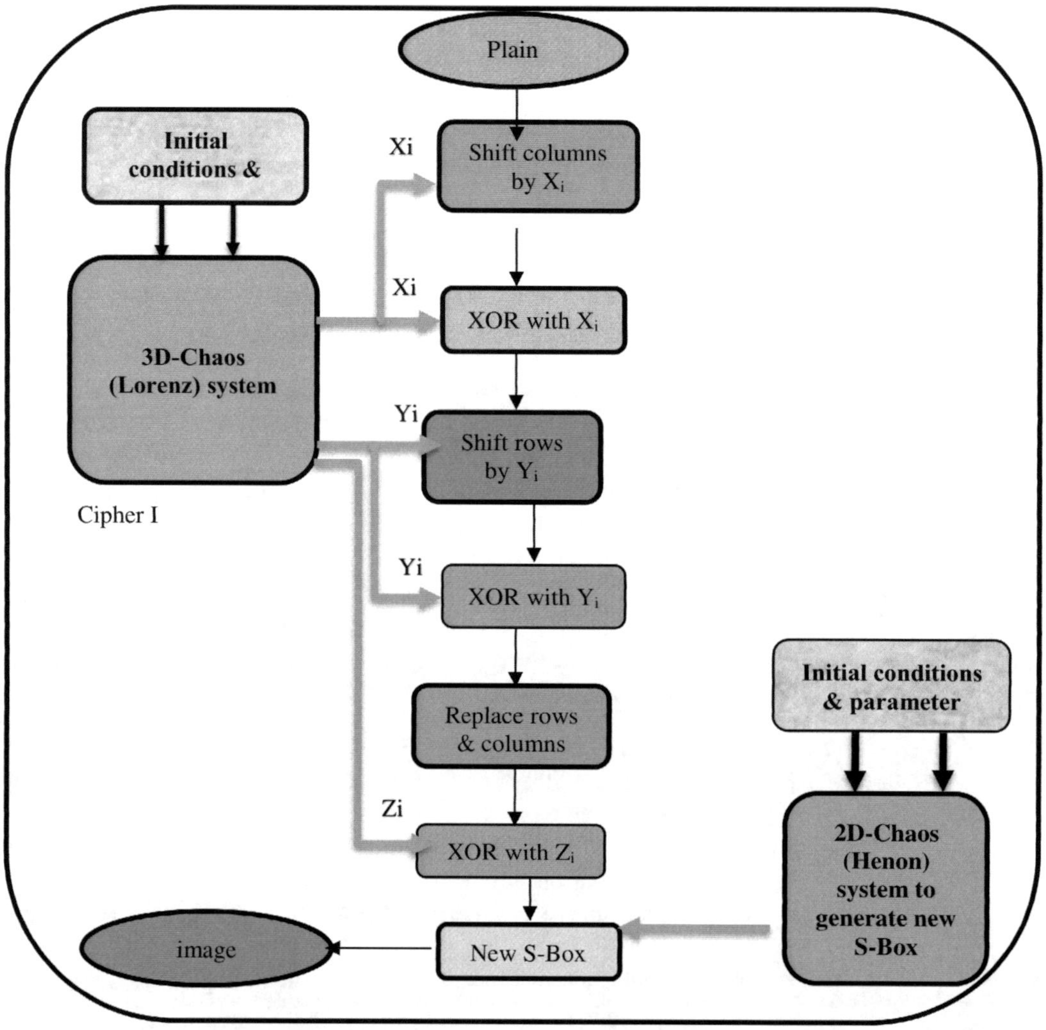

Figure (1): Diagram of Proposal Encryption algorithm

7. Simulation Analysis & Experiment Result

For the purpose of resisting brute-force attacks, the keyspace must be adequately large for securing the image cryptosystem. This segment discusses the results of the suggested encryption algorithm, also the new S-Box for statistical accuracy and analysis of the encrypted image. The base part of the cryptographic block cipher is confusion; each plain-image includes blocks that are transformed into cipher-image blocks, and this builds on the 2D Henon map key for XOR operation. Small changes in initial parameters or conditions lead to various results in final encryption image. Diffusion is the second part of Shift operation in the cryptography block cipher; several digits of the ciphertext can affect every digit of the plain image and every digit of the hidden key.

In terms of the image cryptosystem shown in **Figure 1,** we used the plain images Paper 256*256, Barbra 512*512, Baboon 560*560, Lenna 755*755, Goldhill 900*900 and all are color images and the test results are **as shown in Figure. 2**.

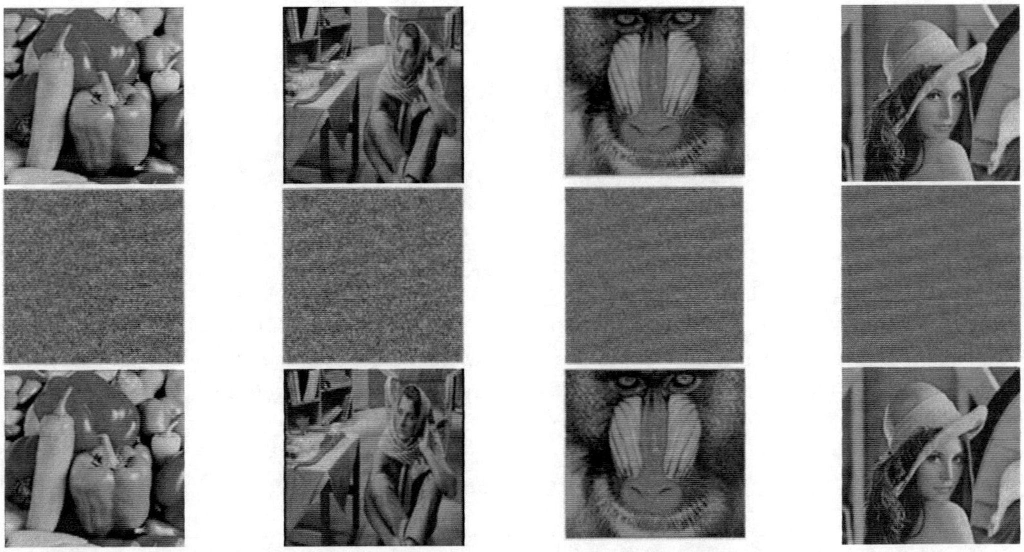

(a) (b) (c) (d)

Fig. 2. Simulation results. (a) Pepper (b) Barbra (c) Baboon (d) Lenna

In our proposed approach, we used different sizes and quality image tests for evaluating the efficiency and security of our proposed system. By Picture Quality Evaluation (PQE) [21], Histogram Analysis Randomness tests and evaluates Image Quality by Entropy.

8. Picture Quality Evaluation (PQE) Metrics

With regard to showing encoded and decoded image quality measurement, the PQE should be utilized as can be seen in the image indicated below, these metrics implemented applied our proposal, **Table 1** show twelve measurements in MSE should be big number because it shows differently between plain image and cipher image with all images show big numbers, the reason **PSNR** results show with these numbers to calculate ratio max probable signal power and noise power, AD show the difference between plain and cipher image and divided by MSE, MD show maximum error between plain and cipher image convert both images to gray image with rang (0-255), NC must be shown in all images 1 between the decryption image and the plain image should be big number because it shows differences between plain image and cipher image, MAE show absolute same idea MSE instead of the square difference between plain and decryption image calculate absolute, NAE show 1 if plain and decryption have no deformation the but the result evaluation show less one, SNR Show all-electric signals between plain and encryption image, SIM show similar results between the original image and encoder image the same idea MSE, and EQ show encryption quality with all images show big results.

Table 1 - PQE Metrics

Name	**MSE**	**PSNR**	**AD**	**MD**	**NC**	**MAE**	**NAE**	**SC**	**NSR**	**SIM**	**CC**	**EQ**
Pepper	10150.39	0.0047	4208.57	240	0.234	8417.59	0.2347	1.2601	2.0640	123.6	0.00568	13266
Barbra	8933.866	0.0053	4679.88	239	0.261	32.4567	0.2616	1.3690	1.9551	122.4	0.00235	14180
Baboon	9810.720	0.0049	4505.42	238	0.250	9011.28	0.2505	1.5090	1.9979	118.9	0.00644	13898
Lenna	9016.81	0.0053	3701.33	211	0.205	7403.39	0.2059	1.0429	2.2034	129.2	0.00263	12609

Encryption & Decryption Running Time

As can be seen in the Table 2, time complexity with all images, time take few milliseconds for encryption and decryption.

Table 2 - Encryption & Decryption Run Time

Name	**Diminution**	**Size**	**Encryption Time**	**Decryption Time**
Pepper	256*256	192KB	953MS	233MS
Barbra	512*512	768KB	100MS	289MS
Baboon	560*560	918KB	455MS	433MS
Lenna	755*755	1.63MB	175MS	867MS

9. S-Box Performance Analysis

Typical statistical criteria, including Balanced Criteria (BC), is the average allocation of 0 and 1 values, Avalanche Criteria (AC) that is an integral criterion that shows how a small shift in the input bits results in the main output change, Strict Avalanche Criteria (SAC) is a condition for each and every cryptographic S-box to state that if an input bit is changed, half of the output bits will be changed, and inevitability, are evaluated according to our proposed approach to designing a new S-box. **(as can be seen in Tables 3 & 4)**

Table 3 - AC, BC comparisons

	AC			**BC**	
	Min.	**Avg.**	**Max.**	**0's Avg.**	**1's Avg.**
Our proposed	0.33	0.570	0.8175	31	32
Ref[22]	0.25	0.5	0.75	32	32
Ref[23]	0.125	0.5	0.875	29	35
Ref[20]	0.25	0.875	0.56	31	33

Table 4 – SAC

	SAC		
	Min.	Avg.	Max.
Our proposed	0.33	0.570	0.8175
Ref[22]	0.25	0.5	0.75
Ref[25]	0.125	0.5	0.875
Ref[24]	0.25	0.875	0.56

10. Differential Attacks Analysis

The Number of Modifying Pixel Rate (NPCR) and Unified Average Adjusted Intensity (UACI) [26,27] were specified as 2 of the major quantities utilized for estimating the strength related to image encryption algorithms/coders for differential attacks, **as shown in Table 5**.

Table 5 - Randomness Tests

	NPSR	**UACI**
Pepper	76595	32747
Barbra	43758	43785
Boats	37489	43785
Lenna	57866	43875
Goldhill	32475	84397

As shown in Table 6, NPSR between our proposed and [28, 29, 30, 31] proposals, and in **Table 7** describe UACI between our proposed and [25, 32, 33, 34] proposals.

Table 6 - NPSR comparisons among different algorithms

	Barbra	**Lenna**
Our proposed	0.987	0.987
Ref[28]	Non	0.994
Ref[29]	Non	0.996
Ref[30]	Non	0.990
Ref[31]	0.996	0.996

Table 7 - UACI comparisons among different algorithms

	Barbra	**Lenna**
Our proposed	0.097	0.865
Ref[25]	Non	0.336
Ref[32]	Non	0.334
Ref[33]	Non	0.335
Ref[34]	0.996	0.335

11. Uniformity Analysis of Image Pixel

The pixel strength diffusion measurements for a picture are represented in a histogram from a picture. A secure encryption system should provide identical histograms to survive statistical attacks. The histogram in **Figure 3** (a, b, c, d) depicts Lena, Pepper, Barbara, Baboon, and Pepper's regular and encrypted pictures. We evaluated from **Figure 3** (a, b, c, d) that the regular image histograms weren't precise, whereas the encrypted digital image histograms have been reliable. The uniformity of the pixel heights of the encrypted image histograms makes it hard to find an insight into the maximum information region for attackers.

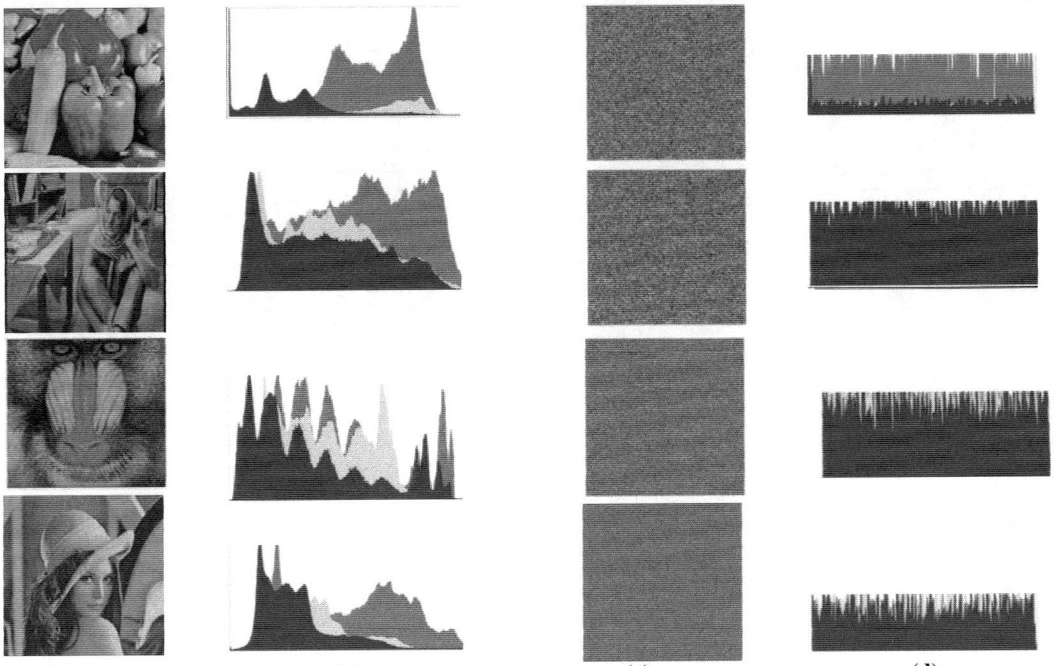

(a) **(b)** **(c)** **(d)**

Figure 3: histogram of Lena, Pepper, Barbara, Baboon, and Pepper (a) plain image, (b) histogram of plain image, (c) cipher image, (d) histogram of cipher image

12. Information Entropy

The entropy of information also is one of the most significant characteristics for calculating the randomness of the cipher file. The H(s) entropy in a source is given by:

$$\mathbf{H(s)} = -\sum_{i=0}^{2^n-1} p(s_i)\log_2 p(s_i)$$

In which, p (s_i) corresponds to the probability related to s. Entropy would preferably be H(s) = 8 for a 2 - 1 gray cipher-8 picture displaying random knowledge, **as shown in Table 8**.

Table 8 - information entropy

Name	**Pepper**	**Barbra**	**Boats**	**Lenna**	**Goldhill**
Inf. Entropy	43646	43645	43546	65654	35656

The entropy in **Table 9** close to the ideal value 8. We thus assume that the algorithm suggested is strongly random.

Table 9 - information entropy comparisons among different algorithms

Image	**Our Proposed**	**Ref[30]**	**Ref[34]**	**Ref[33]**
Lenna	7.9973	7.997	7.997	7.997

13. REFERENCES

[1] Q. Liu, PY. Li, MC. Zhang. YX. Sui, H.J Yang, A novel image encryption algorithm based on chaos maps with Markov pro perties Commun Noilinear Sci. Numer. Simul. 20 (2) (2015) 506-515.

[2] X. Wu, D. Wang, J. Kurths, H. Kan, A novel lossless color image encryption scheme using 2D DWT and 6D hyperchaotic system, Inf. Sci. 349 (2016)

[3] X. Wang, H.L. Zhang, A novel image encryption algorithm based on genetic recombination and hyper-chaotic systems, Nonlinear Dyn. 83 (1-2) (2016) 14021-14033.

[4] Y. Zhang, The image encryption algorithm with plaintext-related shuffling, IETE Tech. Rev. 33 (3) (2016) 310-322. Signal Processing, 147, 2000, pp. 167-175.

[5] Ahmed T., E.M.E.Mostafa, Yasser F. ,Ahmed B. 'Using Chaotic Maps to Enhance RSA Public Key Cryptography', Sci.Int.(Lahore), Vol. 30, PP. 711-715, Sep.2018.

[6] S. Rahma, Abdul Monem , M,Kadhem, Suhad, Engineering and Technology Journal,2012,volum 30,issue 9,pp 1625-1630

[7] Alaa Kadhim F, Hakeem Emad M. 'Mouse Movement with 3D Chaotic Logistic Maps to Generate Random Numbers' Diyala Journal For Pure Science, Vol. 13, PP. 24-39, Apr.2017.

[8] Shokouh Saljoughi, A., & Mirvaziri, H. 'A new method for image encryption by 3D chaotic map', Pattern Analysis and Applications, Vol. 22, PP. 243–257, Nov.2018.

[9] Zhou, N., Pan, S., Cheng, S., &Zhou, Z. 'Image compression–encryption scheme based on hyper-chaotic system and 2D compressive sensing', Optics & Laser Technology, Vol. 82, PP.121–133, Feb.2016.

[10] Hua, Z., Jin, F., Xu, B., & Huang, H. '2D Logistic-Sine-coupling map for image encryption', Signal Processing, Vol. 149, PP. 148–161, Mar.2018.

[11] G. Hatzivasilis, K. Fysarakis, I. Papaefstathiou, and C. Manifavas 2018. A review of lightweight block ciphers, Journal of Cryptographic Engineering., doi: 10.1007/s13389 017-0160-y.

[12] Alaa Kadhim, Rand Mahmoud MohamedAli. Visual cryptography for image depend on RSA & AlGamal algorithms, Conference in 2016 Al-Sadeq International Conference on Multidisciplinary in IT and Communication Science and Applications (AIC-MITCSA).

[13] AT Sadiq, AK Farhan, SA Hassan. A proposal to improve RC4 algorithm based on hybrid chaotic maps, J. Adv. Comput. Sci. Technol. Res, 2016

[14] Alaa Kadhim Farhan, Rasha Subhi Ali, H Rashed Yassein, Nadia Mohammed Ghanim Al-Saidi, Ghassan Hameed Abdul-Majeed,feb. 2020. A NEW APPROACH TO GENERATE MULTI S-

BOXES BASED ON RNA COMPUTING. International Journal of Innovative Computing, Information and Control. Sci., vol. 16, no. 1, pp. 331–348.

[15] Y. Q. Zhang, J. L. Hao, and X. Y. Wang 2020. An Efficient Image Encryption Scheme Based on S-Boxes and Fractional-Order Differential Logistic Map, IEEE Access, doi: 10.1109/ACCESS.2020.2979827.

[16] F. Alaa Kadhim and Z. A. Kamal 2018. Dynamic S-BOX base on primitive polynomial and chaos theory, Int. Iraqi Conf. Eng. Technol. its Appl. IICETA 2018, pp. 7–12, doi: 10.1109/IICETA.2018.8458093.

[17] H. Natiq, N. M. G. Al-Saidi, M. R. M. Said, and A. Kilicman 2018. A new hyperchaotic map and its application for image encryption, Eur. Phys. J. Plus, doi: 10.1140/epjp/i2018-11834-2.

[18] E. N. Lorenz 2017. Deterministic nonperiodic flow, in Universality in Chaos, Second Edition.

[19] A. Kadhim F. and Sura Khalaf. 2015. New Approach for Security Chatting in Real Time, International Journal of Emerging Trends & Technology in Computer Science (IJETTCS), vol. 4, issue. 3, pp. 30–36, doi: 10.24237/djps.1303.268b.

[20] S. El Assad and M. Farajallah 2016. A new chaos-based image encryption system, Signal Process. Image Commun., doi: 10.1016/j.image.2015.10.004.

[21] Mrak, M., Grgic, S., & Grgic, M. 'Picture quality measures in image compression systems'. The IEEE Region 8 EUR OCON. Computer as a Tool, PP. 233-237, Sep.2003.

[22] X. P. Zhang, R. Guo, H. W. Chen, Z. M. Zhao, and J. Y. Wang 2018. Efficient image encryption scheme with synchronous substitution and diffusion based on double S-boxes, Chinese Phys. B, doi: 10.1088/1674-1056/27/8/080701.

[23] I. Yasser, F. Khalifa, M. A. Mohamed, and A. S. Samrah 2020. A New Image Encryption Scheme Based on Hybrid Chaotic Maps, Complexity, doi: 10.1155/2020/9597619.

[24] X. J. Tong 2013. Design of an image encryption scheme based on a multiple chaotic map, Commun. Nonlinear Sci. Numer. Simul., doi: 10.1016/j.cnsns.2012.11.002.

[25] P. Ramasamy, V. Ranganathan, S. Kadry, R. Damaševičius, and T. Blažauskas 2019. An image encryption scheme based on block scrambling, modified zigzag transformation and key generation using Enhanced Logistic-Tent Map, Entropy, doi: 10.3390/e21070656.

[26] J. Khan, J. Ahmad, and S. O. Hwang 2015. An efficient image encryption scheme based on: Henon map, skew tent map and S-Box, doi: 10.1109/ICMSAO.2015.7152261.

[27] Hui L,Bo Z.,Linquan H. 'Quantum Image Encryption Scheme Using Arnold Transform and S-box Scrambling', Entropy 2019, Vol.21, PP.343, Mar.2019.

[28] Essaid, M., Akharraz, I., Saaidi, A., & Mouhib, A. 'A New Image Encryption Scheme Based on Confusion-Diffusion Using an Enhanced Skew Tent Map', Procedia Computer Science, Vol. 127, PP. 539–548, Mar.2018.

[29] Yong Z. 'The unified image encryption algorithm based on chaos and cubic S-Box', Information Sciences, Vol.450, PP.361–377, Mar.2018.

[30] Chai, X. 'An image encryption algorithm based on bit level Brownian motion and new chaotic systems'. Multimedia Tools and Applications, Vol. 76, PP. 1159-1175,Nov.2015

[31] Liu, H., & Jin, C.'A Novel Color Image Encryption Algorithm Based on Quantum Chaos Sequence', 3D Research, Vol. 8, Jan.2017.

[32] Wageda Ibrahim Alsobky ,Abdelkader Esmail ,Ashraf S. Mohra, Ayman Abdelaziem "Design and Implementation of Advanced Encryption Standard by New Substitution Box in Galois Field (2^8)" International Journal of Telecommunications, IJT'2022, Vol.02, Issue 01

[33] Medhat Mansour, Wageeda Elsobky, Ayman Hasan, Wagdy Anis." Appraisal of Multiple AES Modes Behavior using Traditional Enhanced Substitution Boxes ", International Journal of Recent Technology and Engineering (IJRTE) ISSN: 2277-3878, Volume-8 Issue-5, January 2020

[34] Eslam wahba afify, Wageda I. El sobky, Abeer T. Khalil, Reda Abo Alez." Algebraic Construction of Powerful Substitution Box", International Journal of Recent Technology and Engineering (IJRTE) ISSN: 2277-3878, Volume-8 Issue-6, March 2020

[35] Eslam w. afify, Abeer T. Khalil, Wageda I. El sobky, Reda Abo Alez." Performance Analysis of Advanced Encryption Standard (AES) S-boxes ", International Journal of Recent Technology and Engineering (IJRTE) ISSN: 2277-3878, Volume-9, Issue-1, May 2020

[36] Wageda Alsobky, Hala Saeed, Ali N.Elwakeil." Different Types of Attacks on Block Ciphers ", International Journal of Recent Technology and Engineering (IJRTE) ISSN: 2277-3878, Volume-9 Issue-3, September 2020

[37] Wageda I. El Sobky, Ahmed R. Mahmoud, Ashraf S. Mohra, T. El-Garf." Enhancing Hierocrypt-3 performance by modifying its S-Box and modes of operations ", Journal of Communications Vol. 15, No. 12, December 2020

[38] Abdel Halim A. Zikry, Ashraf Y. Hassan, Wageeda I. Shaban, Sahar F. Abdel-Momen.." Performance Analysis of LDPC Decoding Techniques ", International Journal of Recent Technology and Engineering (IJRTEISSN: 2277-3878, Volume-9 Issue-5, January 2021

[39] Mohamed G Abd Elfatah, Hany Nasry Zaky and Ahmed Shams" Mobile Robot Position Estimation using Milstein Algorithm" Journal of Physics: Conference Series, Volume 1970, 10th International Conference on Mathematics and Engineering Physics (ICMEP-10), 7-9 April 2020, Military Technical College, Kobry El-Kobbah, Cairo, Egypt

[40] Mohamed G Abd Elfatah1, Hany Nasry Zaky1 and M Gharib2" Mobile robot position estimation using Euler-Maruyama algorithm" IOP Conference Series: Materials Science and Engineering, Volume 610, 18th International Conference on Aerospace Sciences & Aviation Technology 9–11 April 2019, Military Technical College, Kobry Elkobbah, Cairo, Egypt.

[41] Hany Nasry." Coordinate Transformation In Unmanned Systems Using Clifford Algebra" Proceedings of the 5th International Conference on Mechatronics and Robotics EngineeringFebruary 2019 Pages 167–170https://doi.org/10.1145/3314493.3314496.

[42] Hassan R. Yassein, Nadia M. G. Al-Saidi & Alaa K. Farhan"A new NTRU cryptosystem outperforms three highly secured NTRU-analog systems through an innovational algebraic structure"Journal of Discrete Mathematical Sciences and Cryptography , May 2020

[43] Sameeh Abdulghafour Jassim, Alaa K. Farhan" Designing a Novel Efficient Substitution-Box by Using a Flower Pollination Algorithm and Chaos System" International Journal of Intelligent Engineering and Systems, Vol.15, No.1, 2022.

[44] Jolan Rokan Naif; Ghassan H. Abdul-Majeed; Alaa K. Farhan" Secure IOT System Based on Chaos-Modified Lightweight AES" 2019 International Conference on Advanced Science and Engineering (ICOASE), DOI: 10.1109/ICOASE.2019.8723807

[45] Omar Z. Akif , Sura Mazin Ali , Rasha Subhi Ali , Alaa Kadhim Farhan" A new pseudorandom bits generator based on a 2D-chaotic system and diffusion property" Bulletin of Electrical Engineering and Informatics Vol. 10, No. 3, June 2021, pp. 1580~1588 ISSN: 2302-9285, DOI: 10.11591/eei.v10i3.2610.

[46] F Alaa Kadhim, Ghassan H Abdul-Majeed, Rasha Subhi Ali" Enhancement CAST block algorithm to encrypt big data" 2017 Annual Conference on New Trends in Information &

Communications Technology Applications (NTICT),pages(80-85), **DOI:** 10.1109/NTICT.2017.7976119

A Numerical Study of Decarbonizing Marine Gas Turbine Emissions Through Ammonia/Hydrogen Fuel Blends

Moataz Medhat*, Adel Khalil and Mohamed A Yehia

Department of Mechanical Power, Faculty of Engineering, Cairo University, Giza, Egypt.

*Corresponding author: motaz.medhat@cu.edu.eg

Abstract. The employment of gas turbine in combination with diesel engines and steam generators is a well-known power generation technique in modern marines and ship propulsion. Previously, it rendered its foundations in marine industry through higher power weight ratios and lower NOx emissions if compared to pure diesel engine driven marines. As climate change concerns are becoming more serious, the decarbonization of marine combustion products is becoming of environmental concern. In the present study a modified design of the burner and combustor was suggested to allow for the longer residence time required for releasing the combustion products from the ′slow' burning ammonia molecule. Afterwards, the more formidable challenge of relatively higher NOx emissions was treated through analysis of the effect of altering the equivalence ratio, hydrogen blending, increasing the combustor working pressure and staging the combustion. The latest tactic resulted in lowering values of exit NOx to around 30 ppmv, which is a quite promising result.

Keywards
Gas turbine, CFD, Marine. Ammonia, Hydrogen, Combustion

1. Introduction

The post Covid19 era is changing set of global priorities. The green agenda is becoming of premium status and the elimination of environmental threats is becoming top in agenda of world leaders. About 90% of world's goods are transported by ships. Fossil fuel consumption by marine industry accounts for 330 million metric tons annually most of which are low grade fuels that contributes to a significant portion of emissions of Sulphur Oxide (So_x), Nitrous Oxide (NO_x), Particulate Matter (PM), Carbon Monoxide (CO) as well as Carbon Dioxide (CO_2), [1]. It is the latest that is being considered the main reason for GHG concern at the time being. The main alternatives for fossil fuels that gives no CO_2 in complete combustion are ammonia and hydrogen, [2].

Traditionally, marine propulsion has been powered by two stroke diesel engines with relatively slow to medium speed of typically 800 rpm. However, the continuous power generation featuring gas turbines results in improving the power/weight ratio together occupying less space of the ship enabling higher acceleration and maneuvering characteristics of great need in military ships, allowing addition of more cabins and passengers in cruise ships. Moreover, gas turbine combustion with average temperatures and pressures lower than those attained inside the cylinders of the internal combustion engines leads to lower NO_x emissions and requires much lighter constructions. The steady state nature of combustion of fuels for gas turbines in contrast to that of diesel engines allows constant flow aerodynamic features and well defined chemical kinetics allowing for easier versatility with the type of fuel used and better control on emissions produced.

The combination of both diesel engine and gas turbine in marine industry in different types of assemblies to allow for different conditions differentiate models and types of marines,[3]. Figure 1 shows samples of possible combinations.

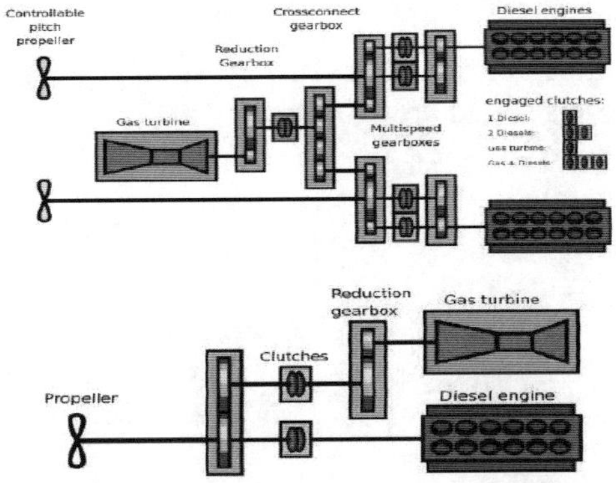

Figure 1. Combined cycles diesel and gas turbines in marine propulsion systems,[3].

Gas turbines are thought to be of superior performance when compared to diesel engines due to their low weight and volume, high speed, reliability, less number of crew on board, easier servicing and maintenance, cheaper maintenance costs and lower emissions,[3]

Predictive studies on marine gas turbine emissions like for example [4] are prompted by the fact that marine pollution negative effects are on the increase. Particulate matter originating from marine combustion and affecting the atmosphere, especially in coastal areas is believed to result in a number of diseases and amounting for around 60,000 mortalities annually,[5]. A bigger concern is the global average temperature increase of approximately 1.0 °C that would reacj 1.5 °C by 2050 due to mainly CO_2 emissions, [2].

The International Maritime Organization set a target on 2018 to reduce GHG from international ships to 50% by 2050. Both hydrogen and ammonia were considered for the solution to the decarbonization problem in ship industry. Ammonia is recently introduced as a carbon free molecule that can function as the renewable clean energy carrier. However, challenges of utilizing ammonia as a power source for fueling naval and generally marine engines includes its high ignition temperature, low heating value, slow laminar flame speed and possibility of forming high NOx content amongst its combustion product. Ammonia can be burnt in both gas turbines and internal combustion engines without producing SO_x or PM together with the elimination of CO_2. Further units have been tested to improve ammonia combustion. Aqueous ammonia burning in HCCI operating mode as anhydrous ammonia with only a slightly higher compression ratio. Aqueous ammonia solutions may be able to improve health and safety concerns for ammonia as well as remove the need for pressurized tanks.

In the present study, a numerical simulation is investigated for two dimensional model of a marine gas turbine combustor via ANSYS fluent solver [6]. The model is validated through comparison of model predictions and previous data [7]. A steady state partial premixed combustion model is selected for predicting emissions and properties resulted from ammonia/hydrogen mixture flame oxidation in a marine combustor model. A new configuration for the combustor is performed to show the enhancement occurring due to the introduced bluff body in the burner geometry. This study elucidates the different combustion characteristics of Ammonia/Hydrogen fuel blend than that of the conventional natural gas, imposing the necessity of different criteria in designing the burner and combustor configurations to yield the aerodynamic pattern suitable for burning the new fuel and providing both high combustion efficiency and low pollutant emissions. Realizable k-ε turbulent model for solving RANS equilibrium equations in steady state condition including the chemistry species mechanism under various conditions. The model studies the influence of various parameters on flame stability and emissions. The study beginning with the influence of equivalence ratios from lean to rich condition on NO_x formation then, the influence of blending hydrogen concentration in the mixture on the flame stability and emissions. Due to the importance of controlling NO_x emissions in marine applications the study present the influence of staging air and inlet pressure on the NO formation.

2. Mathematical Model

The governing equations for 2D, cylindrical coordinates; (r and x) axisymmetric model are solved via ANSYS R20 [6], the general forms stated are as follows:

2.1. Governing Equations

Continuity equation: $\frac{\partial \rho}{\partial t} + \frac{\partial}{\partial xi}(\rho u_i) = 0$ (1)

Momentum equation: $\frac{\partial}{\partial t}(\rho u_i) + \frac{\partial}{\partial x_j}(\rho u_i u_j) = -\frac{\partial P}{\partial x_i} + \frac{\partial}{\partial x_j}(\bar{\bar{\tau}}) - \frac{\partial}{\partial x_j}(\rho\, \overline{u'_i u'_j})$ (2)

Energy equation: $\frac{\partial}{\partial t}(\rho E) + \frac{\partial}{\partial xi}(u_i(\rho E + p)) = \frac{\partial}{\partial xi}(\lambda_{eff}\frac{\partial T}{\partial xi} - \sum_k h_k \vec{J}_k + (\bar{\bar{\tau}}_{eff}\, u_i)) + S_h$ (3)

2.2. Turbulence Modeling

Realizable k- ε model

$$\frac{\partial}{\partial t}(\rho k) + \frac{\partial}{\partial x_j}(\rho k u_j) = \frac{\partial}{\partial x_j}\left[\left(\mu + \frac{\mu_t}{\sigma_k}\right)\frac{\partial k}{\partial x_j}\right] + G_k + G_b - \rho\varepsilon - Y_M + S_k \quad (4)$$

$$\frac{\partial}{\partial t}(\rho\varepsilon) + \frac{\partial}{\partial x_j}(\rho\varepsilon u_j) = \frac{\partial}{\partial x_j}\left[\left(\mu + \frac{\mu_t}{\sigma_\varepsilon}\right)\frac{\partial \varepsilon}{\partial x_j}\right] + \rho C_1 S\varepsilon - \rho C_2\frac{\varepsilon^2}{k+\sqrt{\nu\varepsilon}} + C_{1\varepsilon}\frac{\varepsilon}{k}C_{3\varepsilon}G_b + S_\varepsilon \quad (5)$$

2.3. Combustion Modelling

As discussed before in previous studies [8], [9] Partial premixed model are utilized as a combustion model due to its ability in dealing with fuel mixtures with different reactivity. For predicting the influence of chemical kinetics in turbulent flows using a partially premixed combustion model, the probability density function (PDF) model is selected [10]. To incorporate chemical reactions from the kinetic mechanism [11] into turbulence computations, the following transport conservation equations are used.

$$\frac{\partial}{\partial t}(\rho c) + \nabla.(\rho\vec{v}c) = \nabla.\left(\frac{\mu_t}{S_{ct}}\nabla c\right) + \rho S_c. \quad (6)$$

$$\rho S_c = \rho_u U_t\, |\nabla_c|\ ,\ U_t = A\,(u')^{\frac{3}{4}} S_l^{\frac{1}{2}}\, \alpha^{\frac{-1}{4}}\, l_t^{\frac{-1}{4}} \text{ and } U_t = A\, u'\left(\frac{\tau_t}{\tau_c}\right)^{\frac{1}{2}} \alpha^{\frac{-1}{4}}\, l_t^{\frac{-1}{4}} \quad (7)$$

A is the model constant, the default value given by Zimont [12] is 0.52 and was later reduced to 0.38 following Yehia & Abdel-Raheem [13] , u' is the root mean square velocity, S_l is the laminar flame speed, l_t is turbulence length scale, $\tau_t = l_t/u'$ is the turbulent time scale and $\tau_c = \alpha/\, {S_l}^2$ is the chemical time scale.

3. Problem Description

3.1. Validation

The validation for the CFD model in this work considers the predicted results of previous study El Gohary et al.,[7]. The selection of this study was based upon the relevance of the work studied there to the present study, as it relates to simulating gas turbine combustion in marine applications. The study reported that the gaseous fuels perform well as compared to diesel fuel engines. Although, the engine's compressor and turbines must be changed to handle the variable flow rates in order to accomplish this performance, especially in the case of hydrogen, where considerable changes in flow rates are observed. The 2D axisymmetric combustor is modeled via CFD solver [6]. As shown in Figure 2 the combustion chamber with 1 m long and 0.082 m diameter. Initially, air is introduced directly to the combustor, while fuel is introduced through a nozzle located at the centre of combustion chamber of 0.005 m diameter in a 0.026 m cylinder from the bottom of the burner. The disc employed in the fuel nozzle is responsible for ensuring good mixing in the mixture occurred due to turbulence.

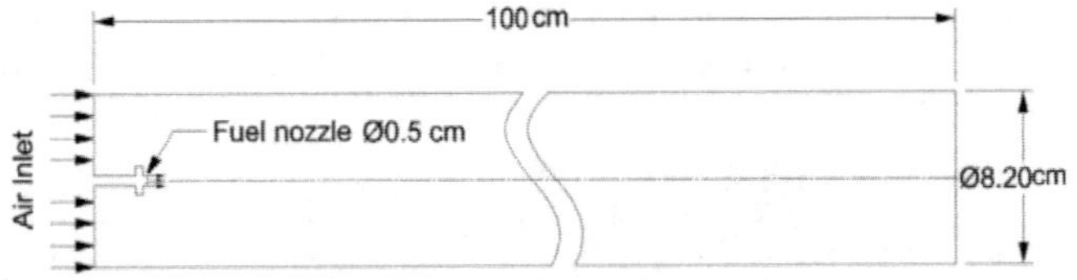

Figure 2. Combustor model, [7].

A 2-D axisymmetric mesh is conducted in this study with quad meshing type of a (0.2 cm) in length as shown in Figure 3. The symmetry in the combustor model has result in reducing the computational cost and time required thus. Only half of the combustor is modelled.

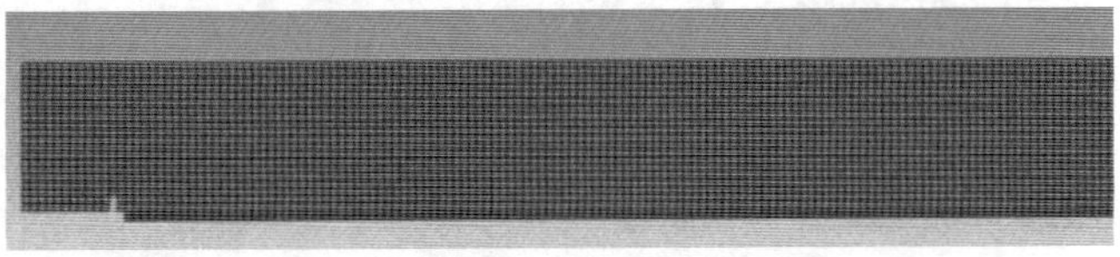

Figure 3. Combustor meshing, [7].

The governing equations for 2D model are solved via ANSYS R20 [6] with steady state case. The turbulence model used is the standard k - ε model and the reactions, solved via a species transport model with eddy concept.

The study [7] compared diesel fuel with natural gas and hydrogen as a gaseous fuels. In this study the validation is conducted only for hydrogen results data. The initial conditions for air and fuel utilized velocities and temperatures at constant pressure. Air introduced in the combustor at 54 m/s with 710 K while, hydrogen entrained from the nozzle at 650 m/s and 300 K. Temperature distribution for the combustor model is validated with the results of the previous data [7] as shown in Figure 4.

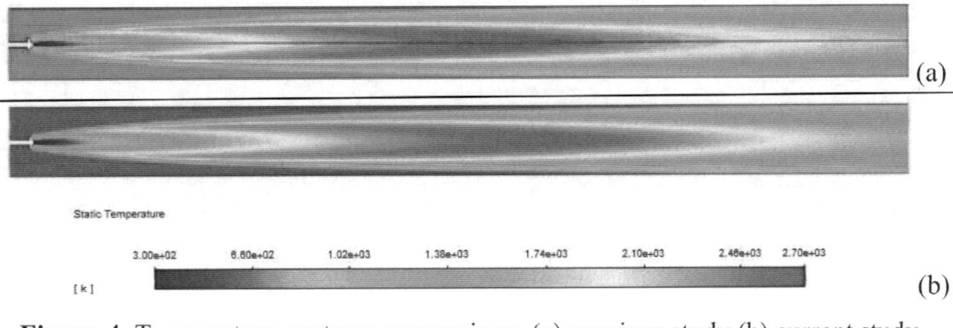

Figure 4. Temperature contours comparison: (a) previous study (b) current study.

On the other hand, oxidation of hydrogen in the combustor introducing high temperature due to the higher reactivity of hydrogen as a fuel results in high NOx formation but, due to its higher flame speed as compared with diesel and natural gas the high temperature zone concentrated in the center of the combustor not in the end of the combustor.

Ammonia is recently introduced as a carbon free molecule that can function as the renewable clean energy carrier. Furthermore, it contains 17.8% hydrogen by weight, with water vapour and nitrogen as the only results of full combustion. In compared to hydrogen, NH_3 has various desirable qualities, such as reduced storage costs, a higher volumetric density that can be easily carried, storage in the liquid phase, manufacture, and a lower reactivity that makes it safer to carry [14]. However, challenges of utilizing ammonia as a power source for fueling naval and generally marine engines includes its high ignition temperature, low heating value, slow laminar flame speed and possibility of forming high NOx content amongst its combustion product. As a result, combining ammonia with other fuels, such as hydrogen, increases reactivity and combustion velocity. Under the same, conditions a mixture of 65% NH_3/35% H_2 burned in the combustor and the results are compared to those of pure hydrogen. Figure 5 shows the differences in the temperature distribution in the combustor when, a mixture of NH_3/H_2 flame is oxidized in the combustor. Maximum temperature is reduced as compared to pure hydrogen due to the low reactivity and lower flame speed of ammonia thus; the high temperature zone is changed in the combustor near to the burner tip.

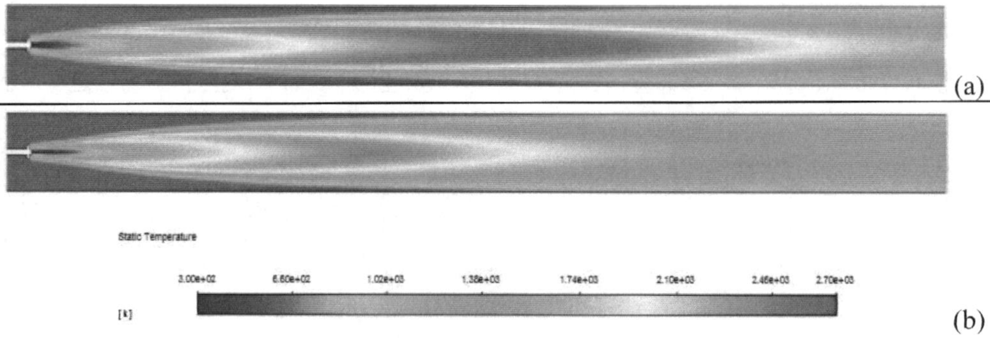

Figure 5. Temperature contours comparison: (a) Hydrogen (b) NH_3/H_2 mixture.

Moreover, the amount of NO_X formation rises due to introducing ammonia in the mixture as shown in Figure 6, when compared with previous data of El Gohary et al.[7]. Modifications in the burner is needed to ensure flame stability and reduce the emissions.

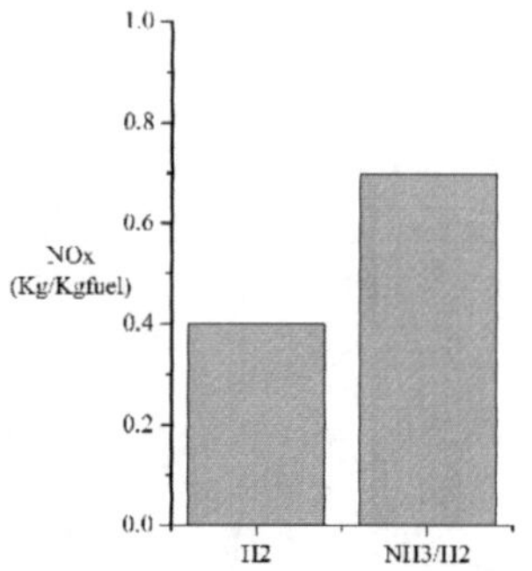

Figure 6. Amount of pollutant per fuel unit mass [kg/kg fuel].

3.2. CFD model

The proposed geometry accumulates the effect of tangentially entrained air and the introduced bluff body for mixing enhancement and flame stability to the mixture in the combustor as discussed previously [8], [15]. The combustor outer dimension is still 1 m long, 8.2 cm in diameter. Although, the air will be introduced through the combustor in cylinder of 8mm in diameter and fuel will issued from two nozzles of 2.5 mm in diameter as shown in Figure 7.

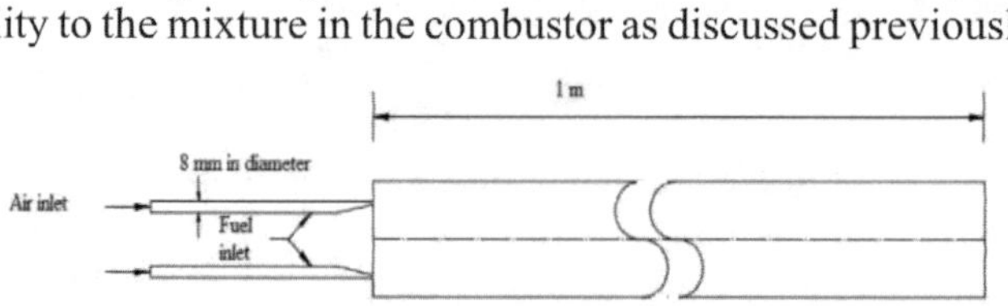

Figure 7. 2D axisymmetric combustor model.

A 2-D axisymmetric mesh of 85000 triangle elements is employed to the geometry as shown in Figure 8. A medium mesh of 3 mm is utilized in combustor but, the grid is fined via 0.5 mm for element in inlet and outlet from the combustor.

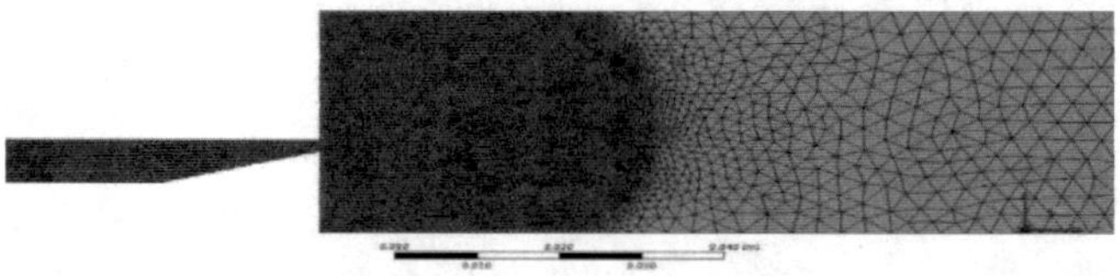

Figure 8. 2-D axisymmetric mesh.

A comparison between three-grid sizes is employed to ensure the independency of the results on the grid size. As shown in Figure 9 axial temperature calculated via CFD software for oxidation of 65% NH_3/H_2 at the three grid sizes. In compared to grid sizes of 85,000 elements and 98,000 elements, which produce nearly equal results, the grid size of 33,000 elements reveals a difference in expected axial temperature. As a result, a grid size of 85,000 elements is deemed appropriate for this study to reduce the computational cost and time.

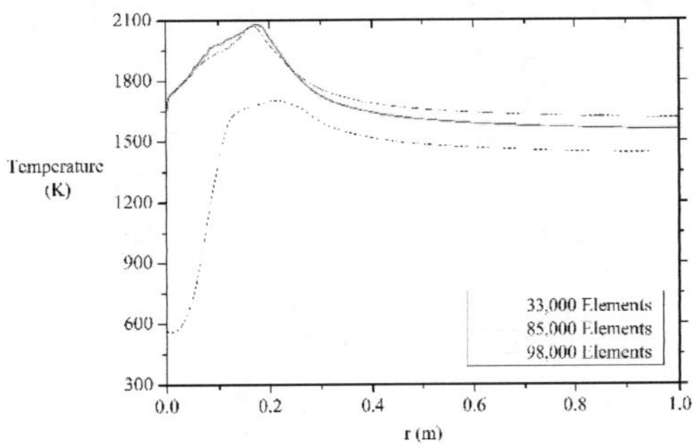

Figure 9. Axial temperature predicated for three different grid sizes for 65% NH_3/H_2 mixture flame.

3.3. *Solution algorithm*
ANSYS software [6] is used to discretize and solve the governing equations. The Coupled algorithm is employed in the iterative solution procedure of the discretized equation. The Coupled approach converges quickly and simultaneously solves the momentum and pressure-based continuity equations. An implicit discretization of pressure gradient terms in the momentum equations, as well as an implicit discretization of the face mass flux, is used to accomplish full implicit coupling. Except for the energy equation for 10^{-6} tolerance, the inner tolerances used to evaluate convergence on the governing equations are set to10^{-3}. The turbulent flow is modelled using the Reynolds Averaged Navier-Stokes (RANS) realisable k- ε scheme with the aid of chemical mechanism [11], which includes laminar flame speed fluctuations with equivalence ratio, as described by Otomo et al.,[9]. Radiation effects are also taken into account, as well as a partially premixed combustion model with flame-let idea.

4. Cases Studied and Boundary Conditions
The model studies the influence of various parameters on flame stability and emissions for 10 cases. The cases simulated via the 2D model beginning with the influence of equivalence ratios from lean to rich condition then, the influence of blending hydrogen concentration in the mixture on the flame stability and emissions. Finally, the study presents the influence of staging air and inlet pressure on the NO formation. Air and fuel initially entrained the combustor at temperature of 400 K and 300 K respectively.

5. Results
5.1. *Effect of equivalence ratio*
The effect of equivalence ratio on the flame stability and emission is studied for a mixture of 65%NH_3/H_2 to map the changes in the flame characteristic from the lean to rich condition. Table 1 illustrates the initial conditions for air and fuel utilized velocities and temperatures at constant pressure for a mixture of NH_3/H_2 burned in the combustor at various equivalence ratios. Velocities are estimated according to the required flow rate for inlet area and air to fuel ratio.

Table 1. Conditions of cases modeled at various equivalence ratios

Cases	Mixture composition	Air		Fuel		Φ	Pressure
		V (m/s)	T (K)	V (m/s)	T (K)		bar
1	65%NH_3/35% H_2	65	400	54	300	0.8	1
2	65%NH_3/35% H_2	52	400	54	300	1	1
3	65%NH_3/35% H_2	46	400	54	300	1.2	1

Figure (10, 11) illustrates the effect of changing equivalence ratios from lean to rich on maximum temperature and NOx formation in the combustion chamber respectively. When the mixture oxidized at rich condition temperature and flame length appeals to be higher than in lean condition. However, the amount of NO formation at the combustor outlet reaches a maximum value at roughly Φ=0.8, while as the equivalence ratio rises, it begins to decline.

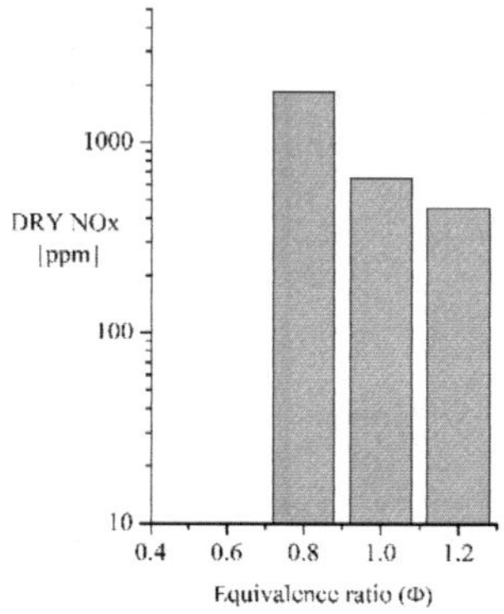

Figure 10. Dry volumetric NO_x formation for various equivalence ratios.

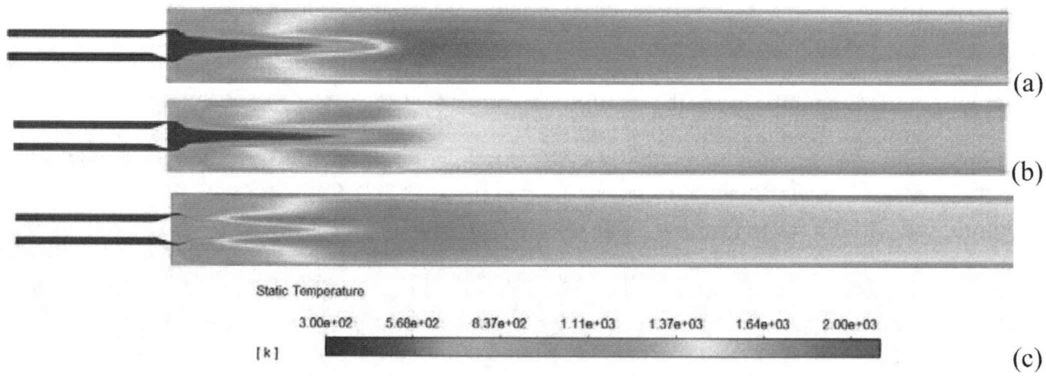

Figure 11. Temperature contours for various equivalence ratios (a=0.8, b=1 and c=1.2).

5.2. Effect of H_2 blending

The study then investigate the influence of blending hydrogen in the mixture on the flame stability and emissions take into consideration the effect of burning the mixture at rich condition, which lead to reduction in NO formation at the combustor outlet. Table 2 illustrates the initial conditions for air and fuel utilized velocities and temperatures.

Table 2. Conditions of cases modeled at various hydrogen blending

Cases	Mixture composition	Air		Fuel		Φ	Pressure
		V (m/s)	T (K)	V (m/s)	T (K)		bar
1	50%NH_3/50% H_2	46	400	44	300	1.2	1
2	65%NH_3/35% H_2	46	400	54	300	1.2	1
3	90%NH_3/10% H_2	46	400	61	300	1.2	1

The bluff body appeals to be an obstacle in the flow stream resulted an internal recirculation zone at the combustion chamber, where the temperature rises, dilution occurs, and the temperature drops until it reaches the outer recirculation zone, as discussed previously by medhat et al.[8]. Figure 12 depicts the influence of the bluff body by introducing two recirculation zones, the first of which is the inner recirculation zone (IRZ), which is primarily related to the introduced swirling flow and bluff-body design, and the second of which is the smaller external recirculation zone (ERZ), which is generated by the rapid expansion of the swirling flow from annuals into the burner, [16]. The improvement of mixture

Journal of Physics: Conference Series **2304** (2022) 012008 doi:10.1088/1742-6596/2304/1/012008

mixing via transporting radicals, unburned, and burned gases into the combustion chamber is achieved as a result of the production of the IRZ in the burner. The IRZ's reduced average velocity resulted in a longer duration for chemical reactions, resulting in relatively greater reaction rates. The temperature distribution shows a rise in temperature near the bottom of the burner as a result of the recirculation zone's influence on increased mixing, indicating that hydrogen concentration is used near the bottom of the burner.

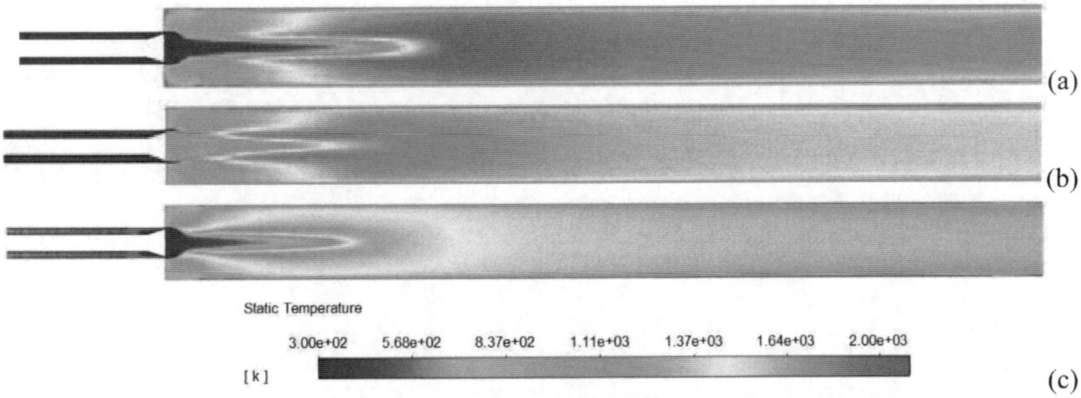

Figure 12. Temperature distributions for different fuel concentrations (a=50%, b=65% and c=90% NH_3) at rich condition Φ=1.2.

Furthermore, the results show that increasing the amount of hydrogen blended in with the ammonia increases the maximum temperature and flame length in the combustor. NO_x formation is then predicted to present the influence of increasing hydrogen blending in the mixture on the NO emission as shown in Figure 13. NO_x formation at the burner outlet proportion increase as a result in increasing in hydrogen concentration.

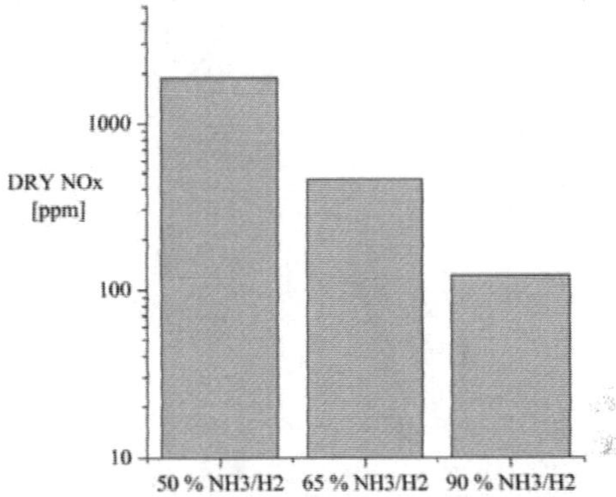

Figure13. Dry volumetric NO_x formation for various concentration of ammonia/hydrogen mixture.

5.3. Effect of secondary air staging

In a gas turbine combustor model, the effect of secondary air designs is investigated in the current study. The air that enters the gas turbine combustor is separated into two streams: main air (combustion air) and secondary air (used for cooling). The effects of secondary air staging on maximum temperature and NOx production in the combustion chamber are shown in Figures (14 and 15). The increase in secondary air causes dilution in the flame of an NH_3/H_2 mixture, resulting in a reduction in the maximum temperature and NOx production.

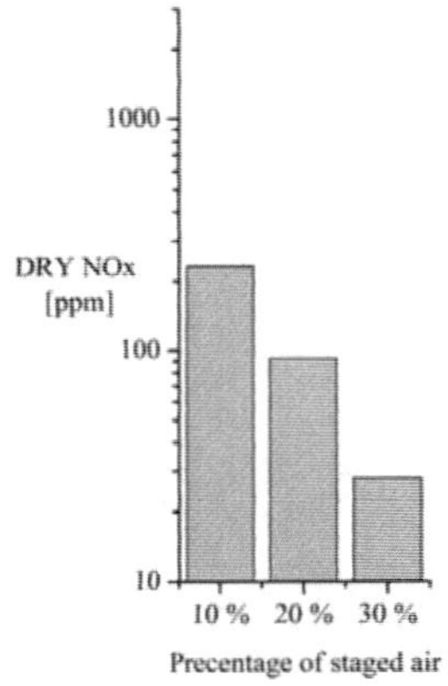

Figure 14. Dry volumetric NOx formation for various amount of staged air for 65%NH3/H2 mixture at rich condition Φ=1.2.

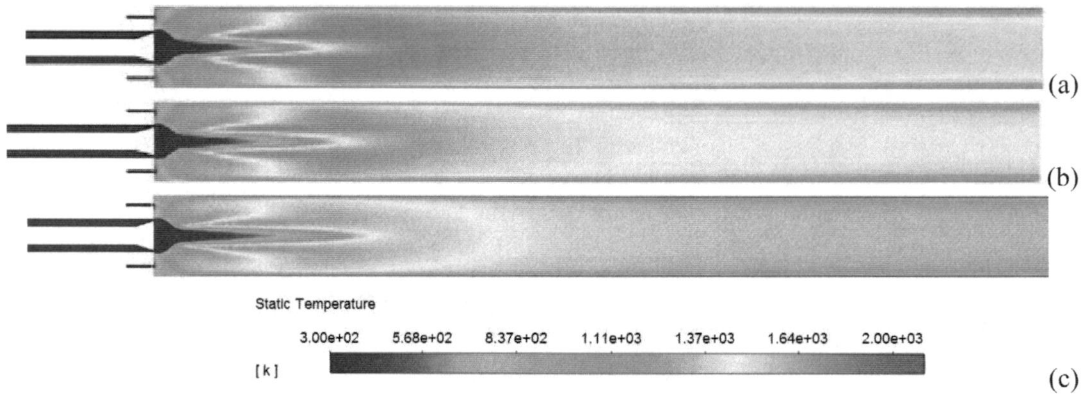

Figure 15. Temperature contours for various amount of staged air (a=10%, b=20% and c=30%) for 65%NH_3/H_2 mixture at rich condition Φ=1.2.

5.4. Effect of Pressure

Figure 16 indicates that increasing the operating mixture pressure to 7 bar above atmospheric pressure has a considerable impact on the decrease of NOx emission concentrations.

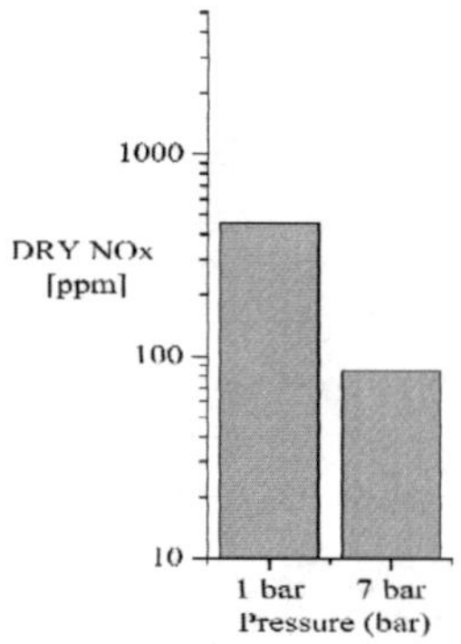

Figure 16. Dry volumetric NOx formation for various operating pressures for 65%NH_3/H_2 mixture at rich condition Φ=1.2.

6. Conclusion

The following is a synopsis of the computational analysis of burning ammonia hydrogen mixtures in a marine gas turbine combustor.

i. The stability of the mixture is achieved by conveying the radicals, unburnt, and burned gases into the combustion chamber as a result of the formation of the IRZ in the burner.
ii. The flame blows off into the chamber when the hydrogen content in the mixture is reduced under lean conditions.
iii. Rich conditions have a significant impact on the reduction of NO and unburnt NH3 emissions, but the correct value of the equivalence ratio under rich conditions must be chosen to achieve the unburnt NH3 and NO emission reduction.
iv. To limit NOx creation, an equivalence ratio of less than 1.2 is preferred.
v. A high hydrogen concentration in the mixture causes the temperature to rise and NO to develop.
vi. Air staging for up to 30% of the combustor appears to be the greatest strategy to limit NO production.
vii. Raising the working mixture pressure to 7 bar has a considerable impact on NO formation decrease.

Reference

[1] Chiong MC, Kang HS, Shaharuddin NMR, Mat S, Quen LK, Ten KH, et al. Challenges and opportunities of marine propulsion with alternative fuels. *Renew Sustain Energy Rev.* 2021;**149** (November 2020):111397. Available from: https://doi.org/10.1016/j.rser.2021.111397

[2] Valera-Medina A, Amer-Hatem F, Azad AK, Dedoussi IC, de Joannon M, Fernandes RX, et al. Review on Ammonia as a Potential Fuel: From Synthesis to Economics. *Energy & Fuels.* 2021;1–108.

[3] Heil G. Marine Gas Turbines. *Mar Eng.* 1980;**85**(10).

[4] Navigation A. Research on Prediction of NO x Emission Performance of Marine. 2021;(x):8–12.

[5] Corbett JJ, Winebrake JJ, Green EH, Kasibhatla P, Eyring V, Lauer A. Mortality from ship emissions: A global assessment. *Environ Sci Technol.* 2007;41(**24**):8512–8.

[6] ANSYS CFX-Solver Theory Guide. 2009;15317(**April**):724–46.

[7] El Gohary MM. Environmental analysis of present and future fuels in 2D simple model marine gas tubines. *Int J Nav Archit Ocean Eng.* 2013;5(**4**):559–68. Available from: http://dx.doi.org/10.2478/IJNAOE-2013-0153

[8] Medhat M, Yehia MA, Khalil A, Miguel C. Franco, Rodolfo C. Rocha. A Numerical Prediction of Stabilized Turbulent Partially Premixed Flames Using Ammonia/Hydrogen Mixture. *J Adv Res Fluid Mech Therm Sci.* 2021;87(**3**):113–33.

[9] Otomo J, Koshi M, Mitsumori T, Iwasaki H, Yamada K. Chemical kinetic modeling of ammonia oxidation with improved reaction mechanism for ammonia/air and ammonia/hydrogen/air combustion. *Int J Hydrogen Energy.* 2018;43(**5**):3004–14. Available from: https://doi.org/10.1016/j.ijhydene.2017.12.066

[10] Vervisch L. Thierry P and Denis V, Theoretical and Numerical Combustion (second ed.), Edwards, USA (2005) 520 pp. *Combust. Flame.* 2006;144(**3**):642–3.

[11] Sun Y, Cait, Shahsavari M, Sun D, Sun X, Zhao D, et al. RANS simulations on combustion and emission characteristics of a premixed NH3/H2 swirling flame with reduced chemical kinetic modeltable 1. *Chinese J Aeronaut.* 2021;(**2021**). Available from: https://doi.org/10.1016/j.cja.2020.11.017

[12] Zimont VL. Gas premixed combustion at high turbulence. Turbulent flame closure combustion model. *Exp Therm Fluid Sci.* 2000;21(**1–3**):179–86.

[13] Yehia MA, Abdel-Raheem MA. Modelling of flammable fuels in small and large scale turbulent environments. *Fuel.* 2020;266(**November 2019**):117110. Available from: https://doi.org/10.1016/j.fuel.2020.117110

[14] Zamfirescu C, Dincer I. Using ammonia as a sustainable fuel. *J Power Sources.* 2008;185(**1**):459–65.

[15] Yehia M, Abdel-Aziz F, Haridy H. Numerical Analyis of Ammonia/Hydrogen Flames in a Swirl and Bluff-Body Stabilized Burner. *Egypt J Eng Sci Technol.* 2020;29(**1**):28–42.

[16] Hayakawa A, Goto T, Mimoto R, Kudo T, Kobayashi H. NO formation/reduction mechanisms of ammonia/air premixed flames at various equivalence ratios and pressures. *Mech Eng J.* 2015;2(**1**):14-00402-14–00402.

Effect of 14.7 MeV Protons on Beryllium Micro Brush / Copper ITER First Wall using SRIM

S I Radwan[1] and H El-Khabeary[1]

[1] Accelerators and Ion Sources Department, Basic Nuclear Science Division, Nuclear Research Center, P.N. 13759, Egyptian Atomic Energy Authority

Email: helkhabeary@yahoo.co.uk
Email: torabsamah@yahoo.com

Abstract. Many simulations carried out on the first wall of ITER. Using SRIM program, the mean distance of 14.7 MeV protons produced from the fusion reaction in ITER which propagated parallel and perpendicular to the beryllium micro brush / copper first wall was calculated. Also, the backscattered ions and vacancies produced per incident proton was calculated. In this study, two simulations were used to show the effect of different numbers of protons incident at normal angle on this target and the other effect was the different incident angles of definite protons number. It was clear that by increasing the number of incident protons, all parameters were not changed except the number of backscattered ions. It was found that with increased the incident protons angle, the lateral projected range increased and reached 1.46 mm at 89° angle. Also, it was found that with increased the incident protons angle, the longitudinal range decreased and both the number of backscattered ions and vacancies have different change.

Keywords: ITER, SRIM Code, Backscattered Ions, Longitudinal Range, Lateral Projected Range and Radial Range.

1. Introduction

European strategy is suggested the 1st trade fusion power plant, DEMO, to follow up from the future project ITER to move fusion energy to the commercial market [1]. Where the **I**nternational **T**hermonuclear **E**xperimental **R**eactor (ITER) under construction at Cadarache - France and is used to generate the electrical power through fusion reaction [2], [3]. It will generate ten times of the input energy about 500 MW of fusion power for periods of time and depend on the magnetic confinement concept of 'Tokamak', in which the plasma is contained in vacuum vessel with a doughnut shape. A mixture of deuterium (D) and tritium (T) is used as the fuel and heated to temperatures in excess of 150×10^6 °C to form a hot plasma. As a result, tritium is not an affluent isotope in nature because of its short life time. For a fusion D-T reactor, it will need to be bred by the nuclear reaction of isotopes as Lithium, Beryllium, or Boron with high energetic neutrons which collide with the first wall. Since several neutrons lost before the creation of T, an effective neutron multiplier is needed to increase the flux of neutron and thus preserving the number of T atoms produced per neutron called tritium breeding ratio (slightly > 1). So that Beryllium (${}^{9}_{4}Be$) is particularly appropriate for this aim as it generates two neutrons instead of the incident one with high probability. The extra neutrons can be used for T production, so the fuel cycle is closed [1]. Then, the small amounts of T liberated when ${}^{9}_{4}Be$ nuclei absorb the neutrons in the three-step nuclear reaction with low energy as following [4]: -

$$^{9}_{4}\mathbf{Be} + ^{1}_{0}\mathbf{n} \rightarrow ^{4}_{2}\mathbf{He} + ^{6}_{2}\mathbf{He}$$

$$^{6}_{2}\mathbf{He} \rightarrow ^{6}_{3}\mathbf{Li} + \boldsymbol{\beta}^{-}$$

$$^{6}_{3}\mathbf{Li} + ^{1}_{0}\mathbf{n} \rightarrow ^{4}_{2}\mathbf{He} + ^{3}_{1}\mathbf{H} \quad (1)$$

Where $\boldsymbol{\beta}^{-}$ is an electron, the half life of $^{6}_{2}\mathbf{He}$ is 0.8 seconds and $^{6}_{3}\mathbf{Li}$ has a high neutrons absorption cross-section.

Since 1994, the first reason for the selection of Be is its acceptable effect on plasma performance and as the first wall due to its erosion lifetime, fuel retention and its performance in ITER [5], [6], [7], [8], [9]. Its life time depends on the erosion and deposition processes and the incidence angle effects were included by an enhancement of self-sputtering yield [10], [11]. To keep the plasma away from the walls, then the strong magnetic fields are used as shown in figure 1.

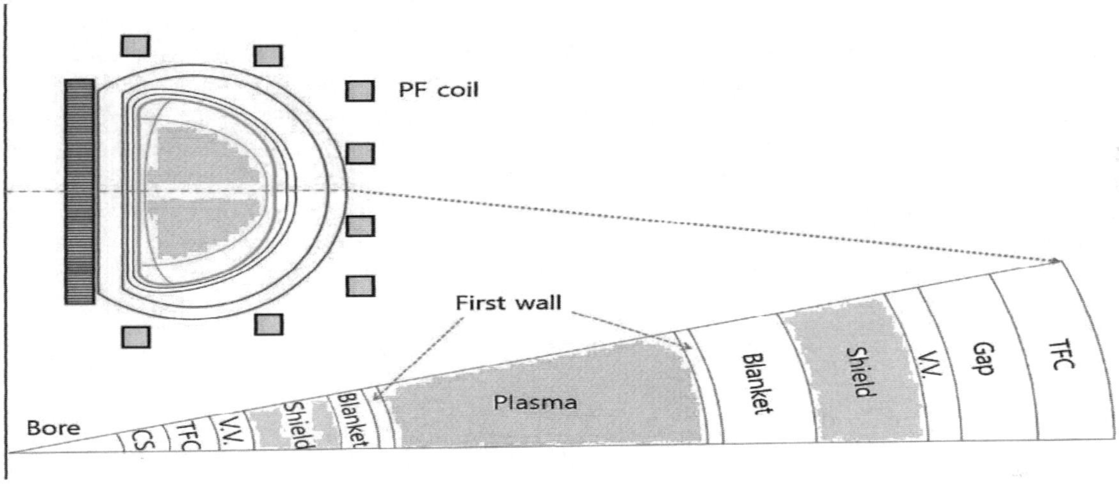

Figure 1. Schematic diagram of ITER design.

Where CS, TFC, PF coil and V.V are central solenoid, toroidal field coils, poloidal field and vacuum vessel. The first easiest nuclear reaction at the lowest energy, is [12], [13]: -

$$^{2}_{1}\mathbf{D} + ^{3}_{1}\mathbf{T} \rightarrow ^{4}_{2}\mathbf{He}\ (\mathbf{3 \cdot 5\ MeV}) + ^{1}_{0}\mathbf{n}\ (\mathbf{14 \cdot 1\ MeV}) \quad (2)$$

This general reaction is a suitable source of neutrons making it to be used in the industrial, research and military applications. The commonly available natural isotope of hydrogen is deuterium and represented 0.02 % of all the naturally occurring hydrogen in the oceans. The easy separation between the hydrogen isotopes is due to the large ratio in mass comparing with the difficult enrichment process of uranium. As a result of short half-life of 12.32 years for the other natural isotope of hydrogen (T) so that it is expensive, hard to find, store and produce. The fusing of two D nuclei is a second easiest fusion reaction with two branches occurring of equal probability:

$$^{2}_{1}\mathbf{D} + ^{2}_{1}\mathbf{D} \rightarrow ^{3}_{2}\mathbf{He}\ (\mathbf{0 \cdot 82\ MeV}) + ^{1}_{0}\mathbf{n}\ (\mathbf{2 \cdot 45\ MeV}) \qquad \mathbf{50\ \%} \quad (3)$$

$$^{2}_{1}\mathbf{D} + ^{2}_{1}\mathbf{D} \rightarrow ^{3}_{1}\mathbf{T}\ (\mathbf{1 \cdot 01\ MeV}) + ^{1}_{1}\mathbf{P}\ (\mathbf{3 \cdot 02\ MeV}) \qquad \mathbf{50\ \%}$$

Where this reaction is also particular in the research field. To initiate this reaction, the optimum energy is equal to 15 keV, slightly higher than that for the D-T reaction (equation (2). The second-generation fusion fuel is:

$$ {}_{1}^{2}\mathbf{D} + {}_{2}^{3}\mathbf{He} \rightarrow {}_{2}^{4}\mathbf{He}\ (\mathbf{3 \cdot 6\ MeV}) + {}_{1}^{1}\mathbf{P}\ (\mathbf{14 \cdot 7\ MeV}) \quad \mathbf{(4)} $$

Where this reaction generates a helium 4 nucleus (^{4}He) meaning α-particles and high energetic protons.

SRIM is a typical code for simulating the damage parameters caused by the ion induced radiation according to the input specifications for its practical usage [14], [15], [16]. Using the Binary Collision Approximation, **BCA**, the profiles of vacancy interstitial Frenkel pairs (FP) as a function of the target depth can be simulated. In SRIM, the first option is Ion distribution and "Quick" calculation of damage using Kinchin-Pease (K-P) model in which the primary ion trajectories are simulated and calculated the produced vacancies. The second one is the full cascade in which after collision by the primary ion, also each recoil atom is followed until its energy falls below the threshold energy of displacement. The process of simulation is continuous whenever the primary ion energy is not sufficient to complete more atom - displacements. Sum of both the average target vacancies and replacement collisions are equal to the average target displacements. It calculates the several physical quantities related to energy deposition, ion sputtering and other effects of interaction between the ions and matter. It allows for various initial positions and the ions' energies including the beam. When the different number of protons travel in material, they lose their kinetic energy due to the interaction with its electrons and nuclei. Here, the expected beryllium depth has been simulated as a result of the interaction of incident protons with the same energy (about 14.7 MeV). It is called the energy loss per unit length and the range, see figure 2. This quantity is described as the longitudinal and lateral ranges which are the mean distance of protons travel parallel and perpendicular to the propagation axis respectively, see figure 3a.

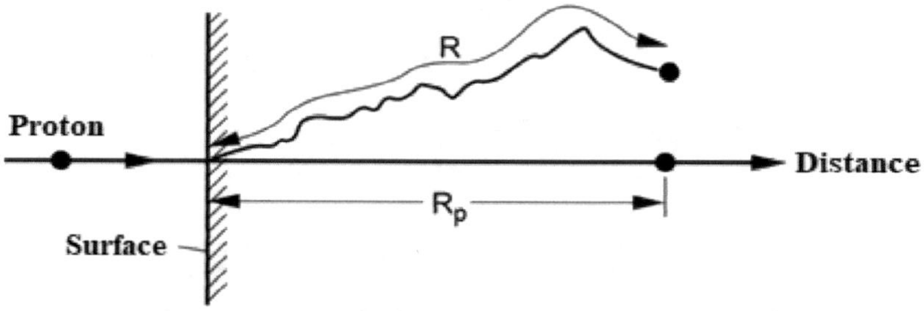

Figure 2. Incident proton on surface with total path length (R) and projected range (R_p) along the direction parallel to the incident proton [17].

As a consequence of stopping and scattering, each individual incident proton forms a random trajectory as shown in figure. 3. Stopping alone defines the total path length Rt. From the endpoint of the trajectory projected ranges can be defined (longitudinal, Rp=, and lateral, Rp⊥). For practical purposes, the (normal) projected range Rp is mostly of interest, since it characterizes the implantation depth with respect to the surface. Obviously, Rp is equal to Rp= for normal incidence. As in most implantation processes the extension of the implanted area is very large compared to the ion range, Rp is mostly also the only range quantity which is accessible to measurement.

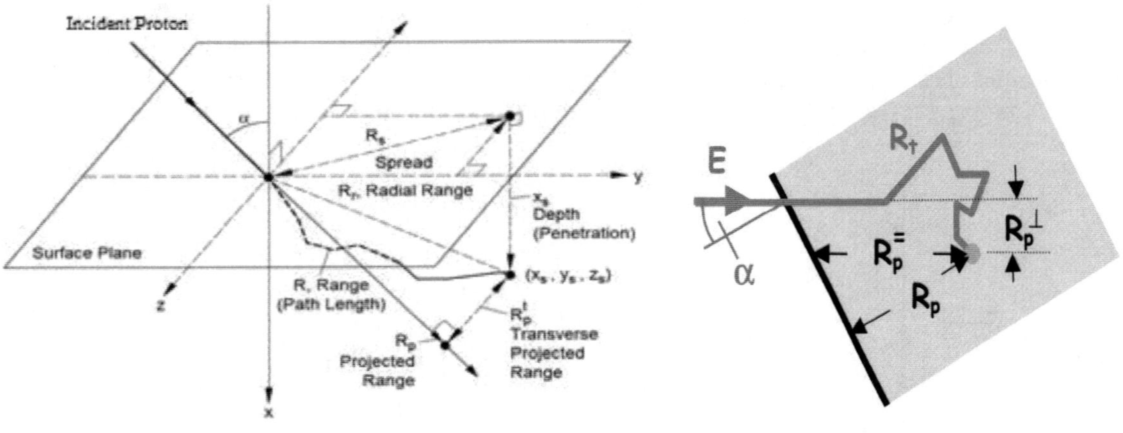

Figure 3. Schematic drawing for the definitions of spread, depth, path length, radial range, transverse projected range and longitudinal projected range [18].

In the previous our studies about the plasma facing components used in future nuclear fusion reactor, we discussed the effect of particles produced from the 1st generation fusion fuel reaction (14.1 MeV neutrons) on some structural fusion materials as graphite, tungsten, 316-stainless steel and copper experimentally using D-T neutron generator with different irradiation time [19], [20]. Then, we simulate the effect of the released particles from the 2nd generation of fusion reaction (3.6 MeV α-particles and 14.7 MeV protons) on fusion materials as graphite, zirconium, titanium, tantalum, molybdenum and tungsten [21].

The aim of this study was studied the effect of 14.7 MeV protons with different numbers incident at angle 89° (maximum angle in SRIM) and also the definite protons number incident at different angles within Be / Cu first wall surface on the parameters produced from SRIM calculations. SRIM simulations were carried out for the target of 100 μm beryllium (Be) layer at the top of target and 5 cm copper (Cu) layer at the bottom of the target plotting at the maximum target depths, as shown in figure 4.

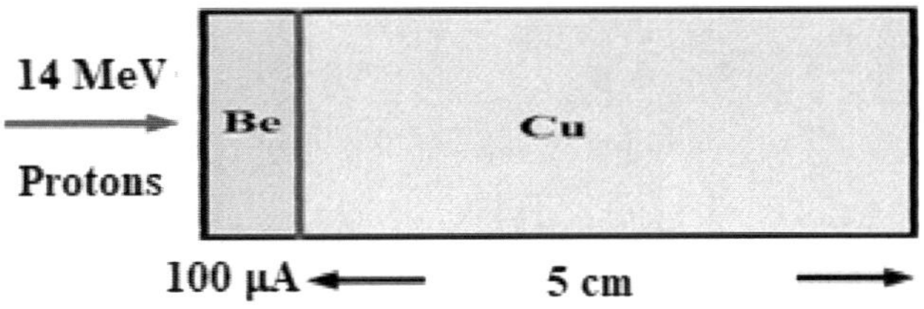

Figure 4. Schematic drawing of SRIM simulation for 14.7 MeV protons incident on micro brush Be / Cu.

2- SRIM Simulations of protons stats on micro brush Be / Cu

The first factor studied was the effect of 14.7 MeV protons with different numbers produced from nuclear fusion reactor on Be micro brush / Cu calculated using SRIM for the total plotting depth at normal angle. Figures 5, 6 and 7 show the effect of different incident protons number at 89° incident angle on the radial range and straggle, lateral projected range and straggle and longitudinal range and straggle, respectively. It was clear from these figures that there was no difference in the radial (range & straggle), lateral projected (range & straggle) and longitudinal (range & straggle) for the different incident numbers of protons because of the same energy. In fact, the mean distance of protons incident parallel or perpendicular on the target was changed by using different protons incident angles.

The projected range is a distance from the surface of implanted material at which the protons reach maximum concentration. It depends on the stay time, the incident angle, the beam current distribution of the incident beam, and the characteristics of the sample. As for the radial range, it depends on the projected range, original ion position, speed and so forth.

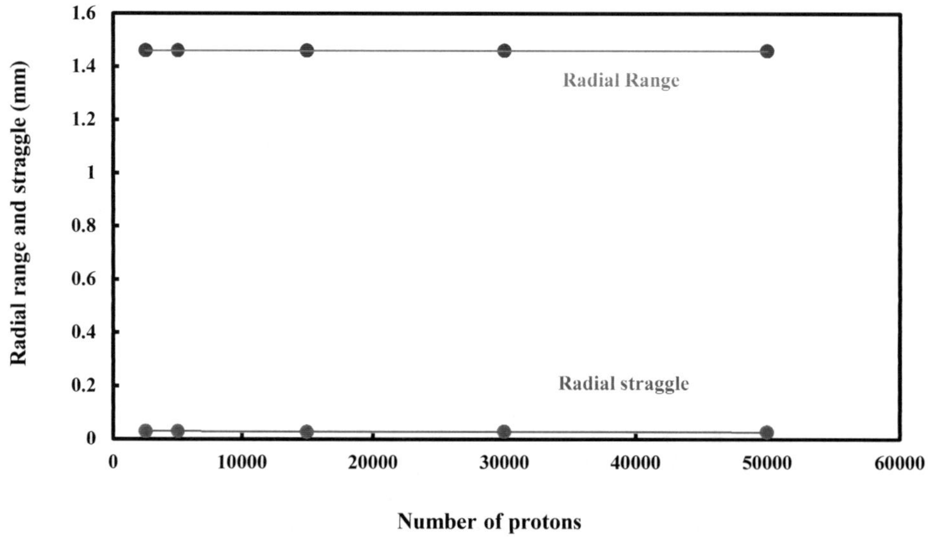

Figure 5. Radial range and straggle (mm) versus different numbers of protons at 89° incident angle on Be / Cu target.

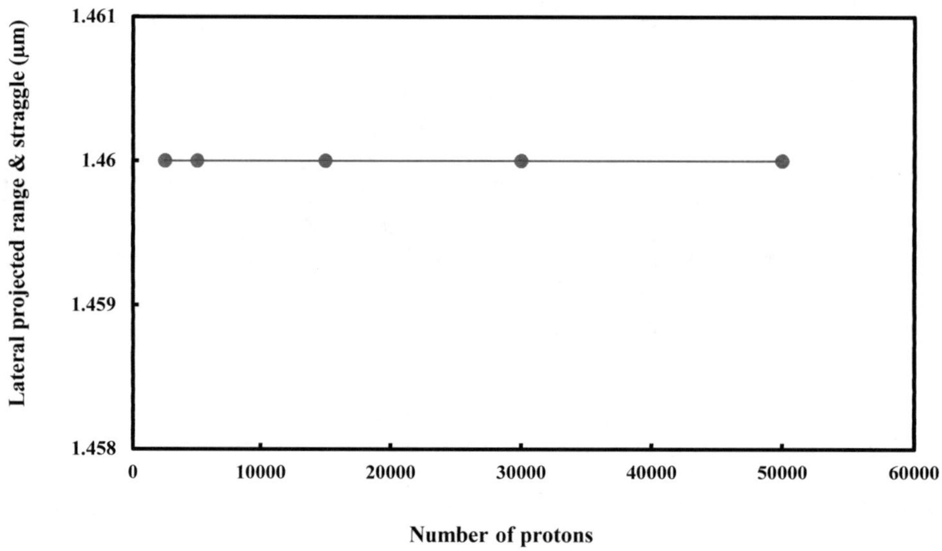

Figure 6. Lateral projected range and straggle (µm) versus different numbers of protons at 89° incident angle on Be / Cu target.

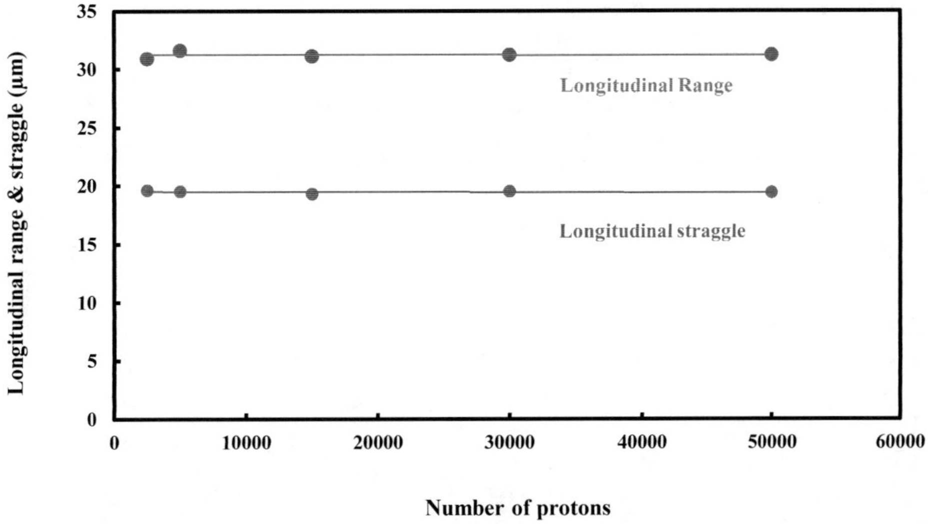

Figure 7. Longitudinal range and straggle (µm) versus different numbers of protons at 89° incident angle on Be / Cu target.

The second factor studied was the effect of definite number of protons (5000) with energy of 14.7 MeV at different incident angles produced from nuclear fusion reactor on Be micro brush / Cu target. Figures 8, 9 and 10 show the effect of 5000 protons of different incident angles equal to 0°, 30°, 45°, 60° and 89° on the radial range and straggle, lateral projected range and straggle, and longitudinal range and straggle, respectively. From figure 8, it was clear that the radial range increased by increasing the incident angle of protons on the target. For radial straggle, it was constant for different incident angles.

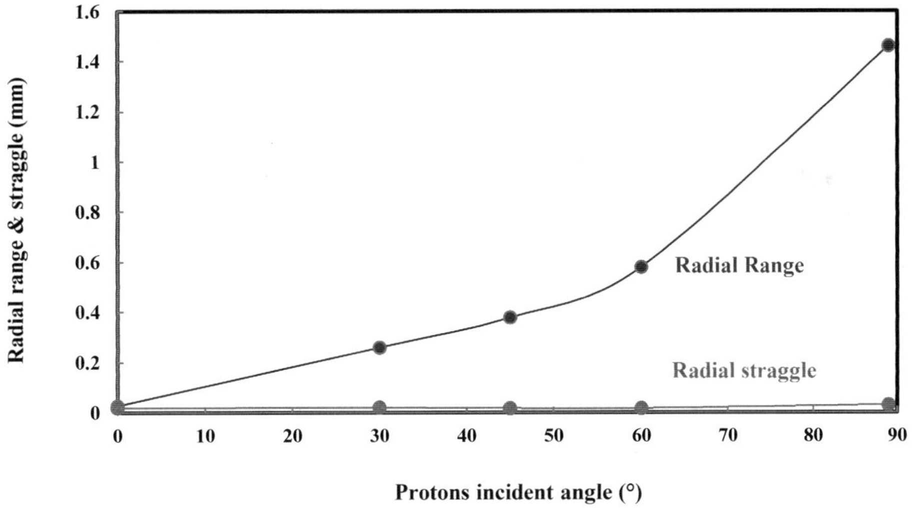

Figure 8. Radial range and straggle (mm) on Be / Cu target versus 5000 protons with different incident angles.

Figures 9 and 10 show the mean distance of incident protons propagated perpendicular to the target axis. It was found that the mean distance of incident protons propagated perpendicular to the target axis increased by increasing the protons incident angle. While the mean distance of the incident protons propagated parallel to the target axis decreased by increasing the protons incident angle.

From figure 9 for many incident protons, the range distribution is smeared out parallel to the surface, with a mean projected range Rp. If the range distribution peaks sufficiently far from the surface, the mean projected range at an angle of incidence is related to the mean projected range at normal incidence. So that the projected range is increased by increasing the incident angle of protons.

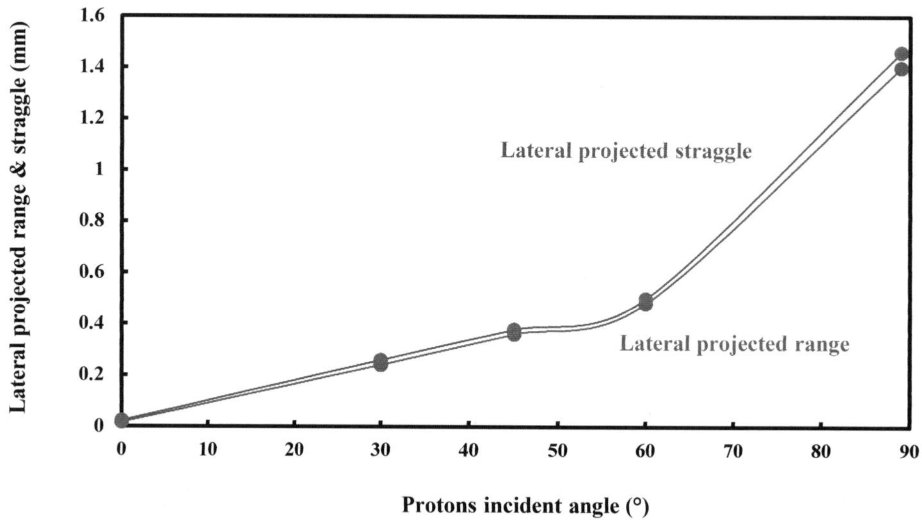

Figure 9. Lateral projected range and straggle (mm) on Be / Cu target versus 5000 protons with different incident angles.

From figure 10 all protons of the same incident energy have the same range. The main parameters governed the range or energy loss rate are the energy and atomic number of the proton Z_1 and atomic number of the substrate Z_2 if we exclude the effect to the orientation of the crystal lattice. it indicates that by decreasing off normal incident angle the protons penetrate to the deeper from the solid surface and vice versa.

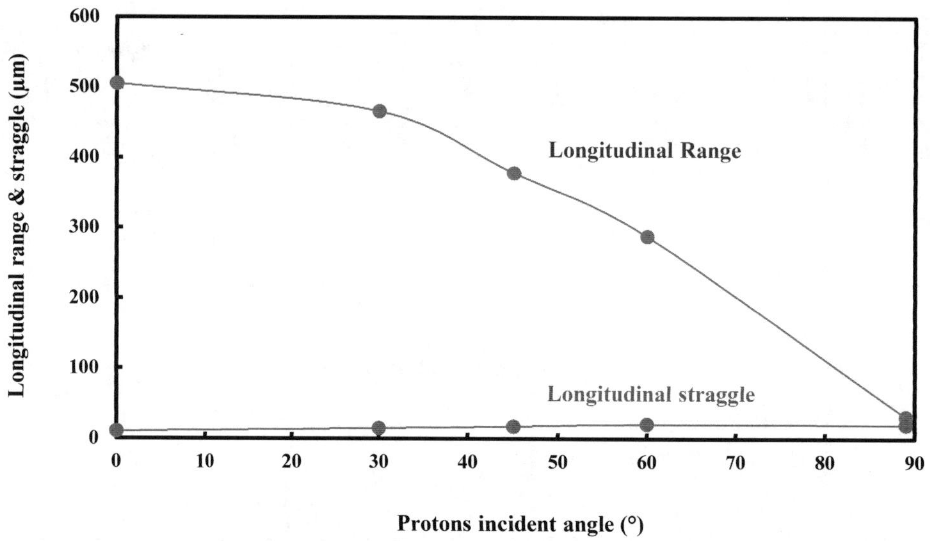

Figure 10. Longitudinal range and straggle (µm) on Be / Cu target versus 5000 protons with different incident angles.

3- Backscattered ions and vacancies diffusion in micro brush Be / Cu

Some ions may be backscattered back out of Be / Cu target surface after strong collision near the target surface calculated by SRIM. Also, the number of vacancies produced due to each incident proton on the target. The first factor studied was the effect of 14.7 MeV protons with different numbers incident at normal angle within Be micro brush / Cu on both backscattered ions and the vacancies produced per incident proton as shown in figures 11 and 12. It was clear that as the number of incident protons increased, the backscattered ions was increased. Also, the number of vacancies changed from 47.1 to 46.3 vacancy per incident proton by increasing the number of incident protons.

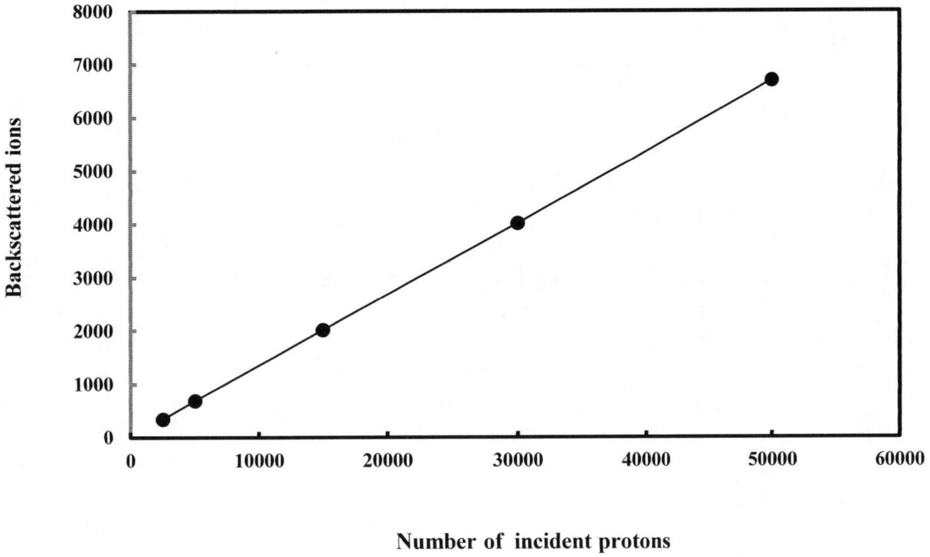

Figure 11. Backscattered ions versus different numbers of protons produced at 89° incident angle on Be / Cu target.

From figure. 11, it was clear that the number of backscattered ions increased with increasing the number of incident protons. Some ions might have a strong collision near the surface of the target and may be backscattered back out of the surface of the target. Backscattered ions are those which, after one or more target collisions, have a trajectory that exits the target back through its surface. From figure 12, it was obvious that very small variation (nearly no change) in the number of vacancies as the number of incident protons was increased. The number of vacancies depends on the displacement Energy assigned to each target atom. A displacement Energy is the minimum amount of energy that must be transferred to the target atom to knock it out of its lattice site and far enough away so it doesn't immediately fall back into this lattice hole. If the transferred energy is less than the displacement Energy, then the atom remains in its lattice site and we assume the energy is transferred into target phonons. If it is greater than the displacement Energy, then the new recoil has this transferred energy, minus the Binding Energy of the atom to its lattice site.

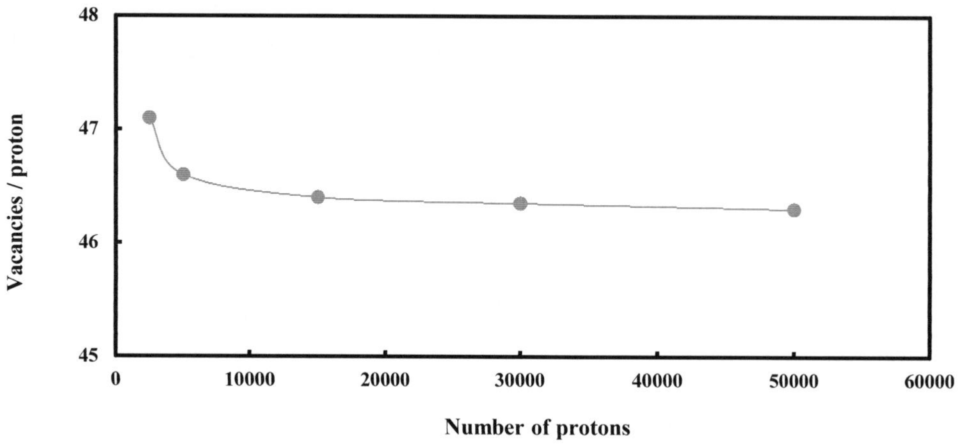

Figure 12. Vacancies produced per proton versus different numbers of protons at 89° incident angle on Be / Cu target.

The second factor studied was the effect of definite number protons (5000) with energy of 14.7 MeV at different incident angles produced from nuclear fusion reactor on Be micro brush / Cu. Figures 13 and 14 show the effect of 5000 protons incident at different angles on the backscattered ions and vacancies per proton produced within Be / Cu target, respectively. From figure 13, it was found that at angle ≤ 60°, the backscattered ions were constant and equal to zero, i.e., there was no backscattered ions. While at normal incident angle equals 89°, the number of backscattered ions was high and reached 681. From this figure, one can observe that the backscattering ions increase substantially as the incident angle changes from θ = 0° (normal incidence) to θ = 89.9° (maximum available grazing incidence).

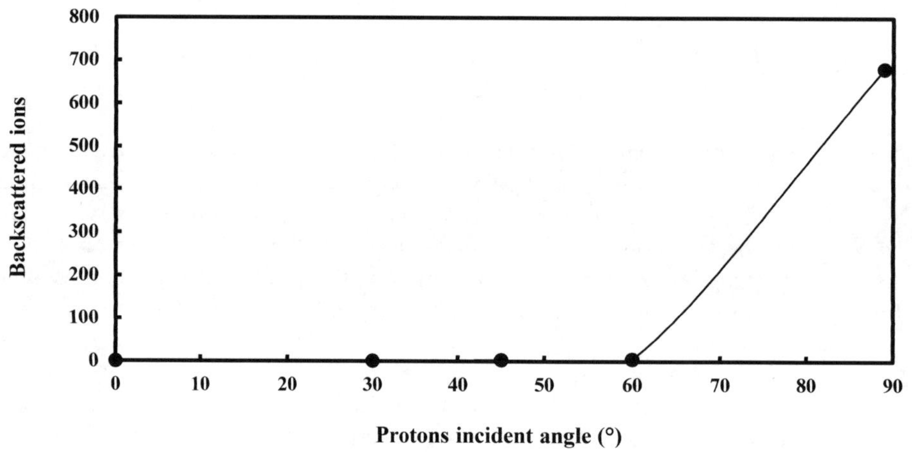

Figure 13. Backscattered ions produced within Be / Cu target versus 5000 protons with different incident angles.

From figure 14, it was clear that the produced vacancies at protons incident angle ≤ 45° was nearly constant (in range about 167.4 - 161.9) and start decrease at protons incident angle equals 60° and reached 46.6 at normal incident angle equals 89°. It was concluded that the value of vacancies produced at protons incident angle equals 0° is nearly 3.6 times of its value at protons incident angle equals 89°. A vacancy is the hole left behind when a recoil atom moves from its original site. So that the number of vacancies is the highest at normal incident angle (θ = 0°) and deceased by increasing the protons incident angle.

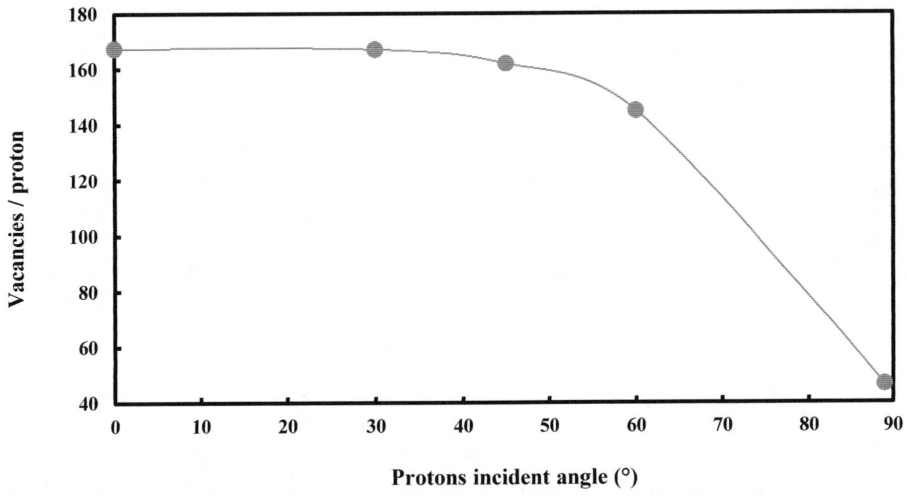

Figure 14. Vacancies produced per proton within Be / Cu target versus 5000 protons with different incident angles.

4. Conclusions

In this study we introduced a behavioral modeling of SRIM for 14.7 MeV protons incident on 100 μm Be / 5 cm Cu as first wall target used in ITER. Here, the simulations were done as a result of two factors to be studied. The first factor was the effect of different numbers of protons incident at normal angle and the second one was the effect of different incident angles of definite protons number at the maximum Be / Cu depths target plotting. Therefore, the longitudinal range / straggle, lateral projected range / straggle, and radial range / straggle were calculated in the two cases in addition to the number of backscattered ions and vacancies per incident proton.

- With respect to the first factor:

i- There was no change in radial (range, straggle), lateral projected (range, straggle) and also longitudinal (range, straggle).
ii- Backscattered ions was increased by increasing the number of incident protons.
iii- Nearly, the number of vacancies decreased by increasing number of incident protons.

- With respect to the second factor:

i- The radial range and lateral projected (range, straggle) were increased by increasing the incident protons angle.
ii- The radial and longitudinal straggle were constant by increasing the incident protons angle.
iii- The longitudinal range was decreased by increasing the incident protons angle.
iv- At proton incident angle ≤ 60, the backscattered ions and number of vacancies were constant and equal to 0 and (167.4 - 161.9) vacancy / proton, respectively.
v- The backscattered ions and number of vacancies were very high and low at normal angle 89°, respectively.

Finally, it was concluded that the different numbers of protons incident on Be micro brush / Cu first wall in ITER at normal angle 89° have no change in the damage depth. But, the damage depth in first wall was changed according to the variation in the incident protons angles rather than normal angle. The different angles except normal angle and the backscattered ions was zero.

References

[1] P. Vladimirov et al, 2014, "Current status of beryllium materials for fusion blanket applications", Fusion Science and Technology, Vol. 66, No. 1, pp. 28-37.

[2] L. Packer et al, 2018, "Activation of ITER materials in JET: nuclear characterisation experiments for the long-term irradiation station", Nuclear Fusion, Vol. 58, No. 9, pp. 096013-096048.

[3] B. Bigot, 2019, "Progress toward ITER's First Plasma", Nuclear Fusion, Vol. 59, No. 11, pp. 112001-112012.

[4] I. Sokolov et al, 2021, "Analysis of the beryllium stability under standard and critical operation in a fusion reactor", Eurasian Journal of Physics and Functional Materials, Vol. 5, No. 4, pp. 188-197.

[5] M. Rubel et al, 2008, "Beryllium plasma-facing components for the ITER-Like Wall Project at JET", Journal of Physics: Conference Series, Vol. 100, No. 6, pp. 062028-062032.

[6] C. Linsmeier et al, 2017, "Development of advanced high heat flux and plasma-facing materials", Nuclear Fusion, Vol. 57, No. 9, pp. 092007-092067.

[7] V. Mukhovatov et al, 2003, "Overview of physics basis for ITER", Plasma Physics and Controlled Fusion, Vol. 45, No. 12A, pp. A235–A252.

[8] V. Barabash et al, 2011, "Summary of beryllium qualification activity for ITER first-wall applications", Physica Scripta, Vol. 2011, No. T145, pp. 014007-014013.

[9] R. Castro et al, 1996, "The structure, properties and performance of plasma-sprayed beryllium for fusion applications", Physica Scripta, Vol. 1996, No. T64, pp. 77-94.

[10] S. Shermukhamedov et al, 2021, "Modelling the sputtering and reflection from a beryllium surface: atomistic analysis", Nuclear Fusion, Vol. 61, No. 8, pp. 086013-086023.

[11] L. Chen et al, 2020, "Sputtering of the beryllium tungsten alloy Be_2W by deuterium atoms: molecular dynamics simulations using machine learned forces", Nuclear Fusion, Vol. 61, No. 1, pp. 016031 -016039.

[12] E. Joffrin et al, 2019, "Overview of the JET preparation for deuterium–tritium operation with the ITER like-wall", Nuclear Fusion, Vol. 59, No. 11, pp. 112021-112046.

[13] P. Batistoni et al, 2012, "Neutronics experiments for uncertainty assessment of tritium breeding in HCPB and HCLL blanket mock-ups irradiated with 14 MeV neutrons", Nuclear Fusion, Vol. 52, No. 8, pp. 083014-083028.

[14] S. Möller et al, 2021, "Scaling of deuterium retention in < 3 MeV proton damaged Beryllium, Eurofer, and W-5Re in the range of 0.0003 to 6 DPA", Physica Scripta, Vol. 96, No. 12, pp. 124051-124056.

[15] A. Lasa et al, 2014, "The effect of beryllium on deuterium implantation in tungsten by atomistic simulations", Nuclear Fusion, Vol. 54, No. 12, pp. 123021-123030.

[16] V. Borodin et al, 2017, "Combined BC/MD approach to the evaluation of damage from fast neutrons and its implementation for beryllium irradiation in a fusion reactor", Modelling and Simulation in Materials Science and Engineering, Vol. 25, No. 8, pp. 084005-084022.

[17] M. Mayer et al, 2020, "Ion beam analysis of fusion plasma-facing materials and components: facilities and research challenges", Nuclear Fusion, Vol. 60, No. 2, pp. 025001-025021.

[18] R. Kwiatkowski et al, 2014, "Computer simulation of charged fusion-product trajectories and detection efficiency expected for future experiments within the COMPASS tokamak", Physica Scripta, Vol. 2014, No. T161, pp. 014013-014018.

[19] S. Radwan et al, 2015, "Simulation of particle impact on protective coating of high-level waste storage packages", Progress in Nuclear Energy, Vol. 81, No. 2, pp. 196-202.

[20] S. Radwan et al, 2021, "Influence of high-energy particles on copper and stainless steel in fusion reactor materials", Vol. 99, No. 11, Canadian Journal of Physics, pp. 954–968.

[21] S. Radwan et al, 2020, "Effect of 14.7-MeV Protons and 3.6-MeV Alpha Particles on Fusion Structural Materials", Fusion Science and Technology, Vol. 76, No. 4, pp. 710-722.

Simulation of Ozone Concentrations Inside Co-60 Industrial Irradiator Using UV Optical Cell

Amal El gamel [1] Ali Noaman [2]

[1,2]Egyptian Atomic Energy Authority, Cairo, Egypt
[1]Nuclear and Radiological Safety Research center& Department of Radiation safety.

[2] National Center for Radiation Research and Technology & Department of Radiation

Email address: [1]amal.gamel@yahoo.com
alino3man2004@gmail.com

Abstract
ozone concentration in the Co-60 industrial irradiator is controlled by the local exhaust ventilation (LEV). if the event of a ventilation malfunction, the ozone concentration becomes increasingly sensible within the radiation chamber. Ozone (O_3) is a toxic gas which can cause tract, asthma, and permanent lung damage. its detrimental effects on humans and corrosion of metal parts within the radiation chamber indicates its importance as a safety factor consideration. The motivation for this work designs a standby system to reduce the ozone concentration within the Co-60 industrial radiator. MATLAB was used to simulate the effect of ultraviolet (254 nm) absorption cells on ozone concentration and calculate the decay time. The simulation illustrated that UV absorption is linearly increasing with increasing ozone concentration. Employing a UV Cell proved to reduce ozone concentrations and their decay time with an efficiency of 99%. This may prevent or significantly reduce the discharge of ozone into the air from the workplace.so it's recommended that's using UV (254 nm) in the ozone concentration within the Co-60 industrial as it is cheap and efficient controlling substance.

Keywords

UV lamp, MATLAB, Lambert-Beers law, ozone decomposition

1. Introduction

In the Co60 industrial irradiator, the air layer between the source and the products under processing interacts with gamma radiation and produces ozone gas. [1] Ozone is a powerful oxidizing agent that can react explosively with oil and grease. Low concentrations of ozone have a significant effect on textiles, fabrics, organic dyes, metals, plastics, and paints, and cause the characteristic cracking of stressed rubber, commonly called 'weathering'. Since ozone is a highly reactive substance, any adverse health effects will be found at the sites of initial contact: the respiratory tract (nose, throat, and airways), the lungs, and at higher concentrations, the eyes, and damage to the small airways of the lungs. To avoid any potential harm to both workers and metallic parts inside the irradiation room, a ventilation system is installed to reduce and control the concentration and release of ozone in the environment within the permitted levels. Effective

risk management measures are established by Control of Substances Hazardous to Health Regulations (COSHH), which ensures that the exposure of their employees to hazardous substances is either prevented or mitigated, is used as reference to control the exposure of ozone. a focus on controlling the exposure to levels at which any health effects are removed. The Control of Substances
The workplace exposure limit WEL [2] for ozone is 0.2 ppm in air averaged over a 15-minute reference period. If exposure to ozone cannot be prevented, then the principles of good control practice need to be applied to ensure that the workplace exposure limit is not exceeded. When the motor fails and the flow rate in the system reaches zero the calculated ozone concentration value excessed the WEL. In the event of failure in the motor responsible for blowing air out of the radiation chamber there must be another control system to reduce ozone concentration inside the irradiation room as soon as possible.
Short-wave UV-C (280 nm) is completely absorbed by ozone [3]. This third oxygen atom is broken by UV light with a wavelength of 240-315 nm. As a result, a UV-C lamp with a wavelength of 253.7 nm will actually destroy ozone.[4] The absorption of UV-C light has long been absorbed by ozone with high precision and accuracy. The ozone molecule has an absorption maximum at 254 nm, the mechanism for removing ozone is dissociation, which occurs when 254 nm UV energy "breaks" one of the oxygen bonds in an ozone molecule. As a result of this reaction, each ozone molecule is converted into one oxygen atom and one oxygen molecule. Free oxygen atoms will combine with each other to form oxygen molecules. as shown in figure (1), which can be critical for process control. There are several gas molecules that absorb strongly in ultraviolet (UV) (wavelength range, 180 nm to 400 nm) and deep-UV (wavelength range, 180 nm to 280 nm) thus, by exploiting the absorption properties of these compounds. a sensitive and selective sensing platform can be devolved as shown in figure (2).
The absorbance of each gas is normalized by the respective maximum absorption coefficient of the gas. The data of each molecule were obtained from HITRAN [6] and then plotted on a wavelength scale.
Common UV sources are open arcs, fluorescent and incandescent sources, lasers, and LEDs. UV sources are characterized using wavelength range, optical output stability, lifetime, and input power, a UV emission source should be stable and have enough power.[5]

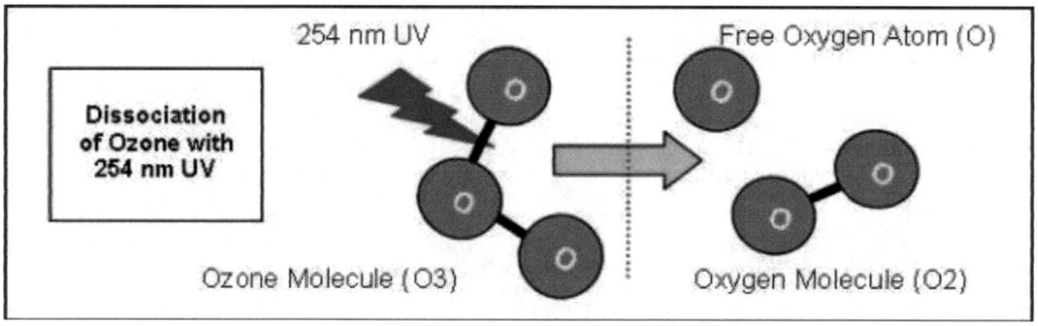

Figure (1) UV Ozone Destruct (Removal).

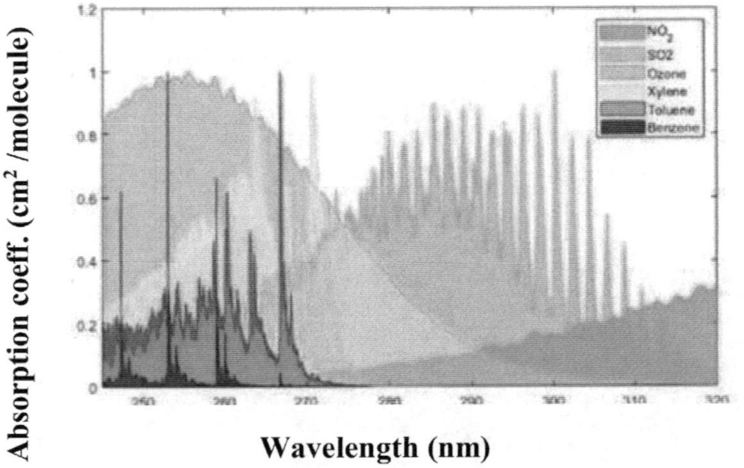

Molecules	Maximum absorption coeff.
Benzene	7.730×10^{-18}
Toluene	1.216×10^{-18}
Xylene(m-)	1.169×10^{-18}
Ozone	1.166×10^{-17}
SO_2	1.227×10^{-18}
NO_2	8.029×10^{-19}

Figure (2). The ultraviolet (UV) absorption spectrum of different gases.

2. Materials and methods

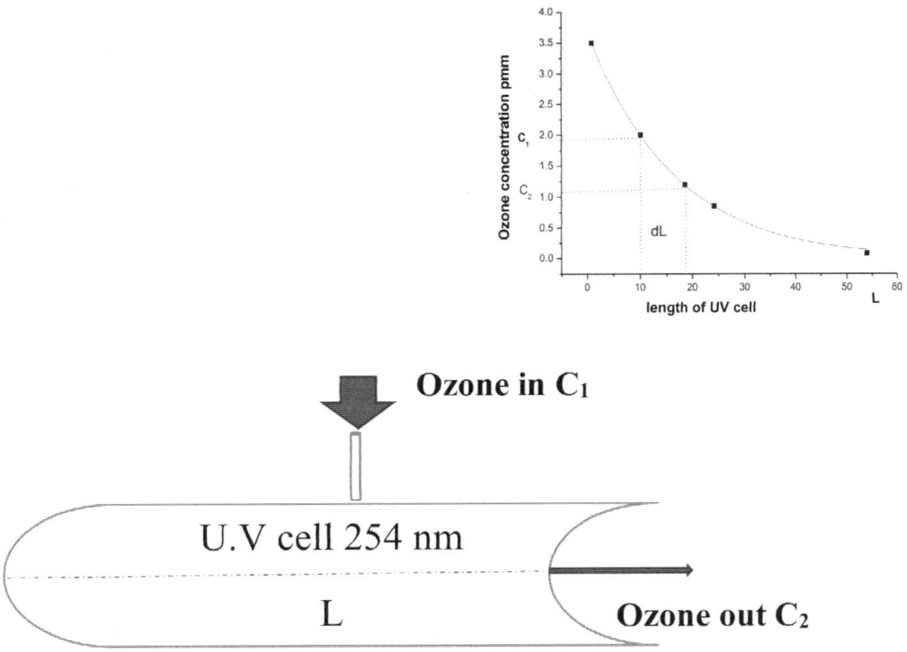

Figure (3) geometry of UV cell in MATLAB simulation

2.1 MATLAB simulation

The analytical principle is predicated on the absorption of UV radiation by the ozone molecule and measuring the reduction of the quanta of light at 254 nm, as shown in Figure (3). The ozone concentration for a UV cell with different pass lengths of 50 cm and 15 cm was determined .The reduction depends on the path length of the UV cell, the ozone concentration introduced into the UV light cell, where Radiation attenuation of UV is defined by equation related to each other according to the Lambert Beer law [7].where the intensity of UV radiation is known as Io. therefore, the intensity of UV is attenuated with the concentration of ozone I, as shown with the equation: (1). Absorption is usually quantified using transmittance and absorbance.

$$I = I_0\, e^{-\sigma l C} \quad (1)$$

$$C = \frac{1}{l\sigma} \ln\frac{I_o}{I} \quad (2)$$

$$A = \ln\frac{I_o}{I} \quad (3)$$

Where;
I_o = the light intensity in the absence of ozone
I = the light intensity in the presence of ozone
C= ozone concentration (molecules/cm3)
A = Absorbance
σ = cross section for ozone at 254 nm = 1.15×10^{-17} cm^2 $molecule^{-1}$
l = the path length of UV cell in cm
we get the formula to calculate the ozone concentration in the ppm

$$C = \frac{RT\,10^6}{\sigma P l N_A} \ln\frac{I_o}{I} \quad (4)$$

Transmittance as given by T_r =I/Io, T affects the sensitivity of the UV cell. and was validated numerically as follows [7];

$$T_r = \exp\left(\frac{-\sigma p l C N_A}{RT10^6}\right) \quad (5)$$

Where , C is the concentration of an analyte in ppm , σ is the absorption cross-section in m^2 $molecule^{-1}$,T is the absolute temperature in K , P is the pressure in atm, R is the ideal gas constant, $8.205746\text{x}10^5$ atm-m^3 Mol^{-1} K^{-1},N_A is Avogadro's constant, $6.02214199\text{x}10^{23}$ molecule-mol^{-1}. Transmittance T affects the sensitivity of the UV cell [8]. derived equations based on Beer–Lambert law for optical path length, the optimized optical path length, l_{op} to calculate output ozone concentrations C_2 were expressed as,

$$l_{op} = \frac{RT(lnC_1 - lnC_2)10^6}{\sigma P N_A (C_1 - C_2)} \quad (6)$$

3. RESULTS AND DISCUSSION

Table 1: shows the measured ozone concentration inside the irradtior room with the exhaust on , the ozone concentration with UV-C (15cm) and the ozone concentration with UV-C (50cm).Figure (4) shows the comparison between the measured values of O3 concentration within the pipe with source 1 Ci using the ozone monogram measurement method and the calculated values with flow 50 m3/ min and the simulation using MATLAB, to check the validity of the software program. Figure (5) shows the ozone decay curve within the pipe, with the exhaust on; V1 and after gathering sources of cobalt-60. It might be seen that after 50 min, the O3 concentration is already less than the limit for workers (0.075ppm). Figure (6) shows the relationship between the measured ozone concentration and the absorbance of ozone using the UV cell. The slope indicates that the absorbance of ozone increases. In addition, a path length of 50 cm is better than a path length of 15 cm for the absorption of ozone. From figure (7) observed that the transmittance of the UV cell Which is meaning that the output ozone from the UV cell is decreasing. Figure (8) shows the relationship between the ozone concentration without and with the UV cell with a path length of 15 cm. and Figure (9) shows an equivalent relation with a path length of 50 cm. Figure (10) shows the ozone decay

Journal of Physics: Conference Series **2304** (2022) 012010 doi:10.1088/1742-6596/2304/1/012010

curve using the UV cell (15cm), with the exhaust working and activity 1 Ci and V3 with flow of 0 m3/ min. it can be seen that after 0.01 min, the ozone concentration is already lower than the limit for workers. Figure (11) shows the ozone decay curve using a UV cell (50 cm) activity 1 Ci and V3 with flow 0 m3/ min. it can be seen that the ozone concentration is already lower than the limit for workers. As shown Figure (12) the efficiency of using a UV-C cell is 99.9 % of that using a UV cell.

Table 1: the simulated ozone concentration ppm

Ozone concentration ppm	Ozone concentration ppm With UV-C cell (15cm)	Ozone concentration ppm With UV-C cell (50cm)
3.5	0.00013	1.10362E-13
2	0.006009	3.94364E-08
1.2	0.046479	3.61026E-05
0.85	0.113757	0.000713274

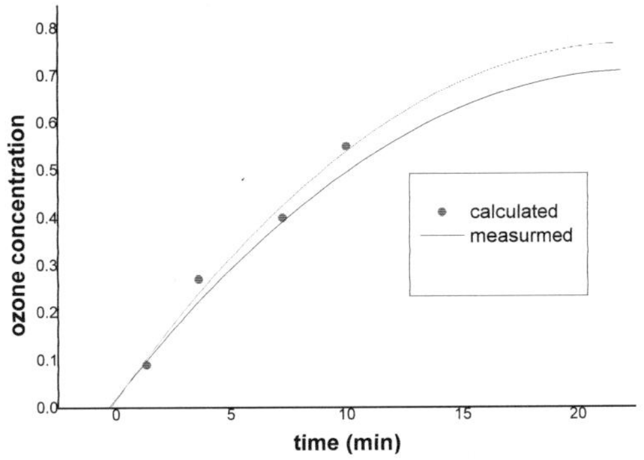

Figure (4) Comparison the monogram measurement method measured and the calculated Value of O_3 concentration in the exhaust pipe with source 1 Ci with Flow rate 50 m^3/min

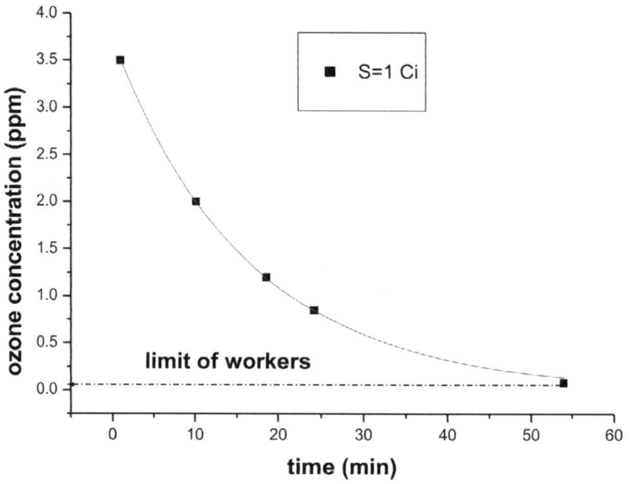

Fig. (5) ozone decay curve in the exhaust pipe, with the exhaust system is not working, in the Co60 industrial irradiator with activity source 1 Ci&V1.

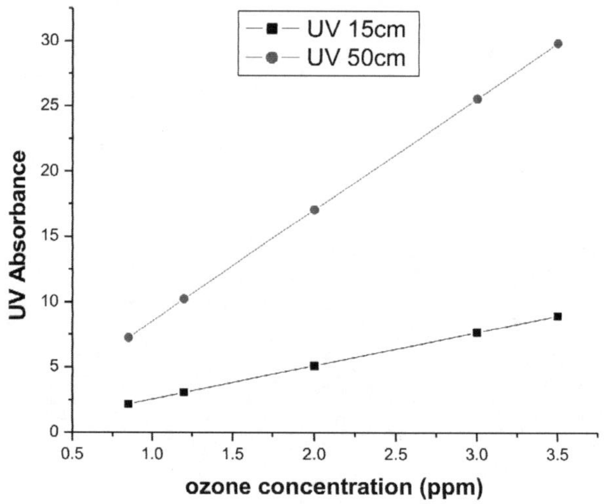

Figure (6) UV cell absorbance with different path length as a function of ozone concentration.

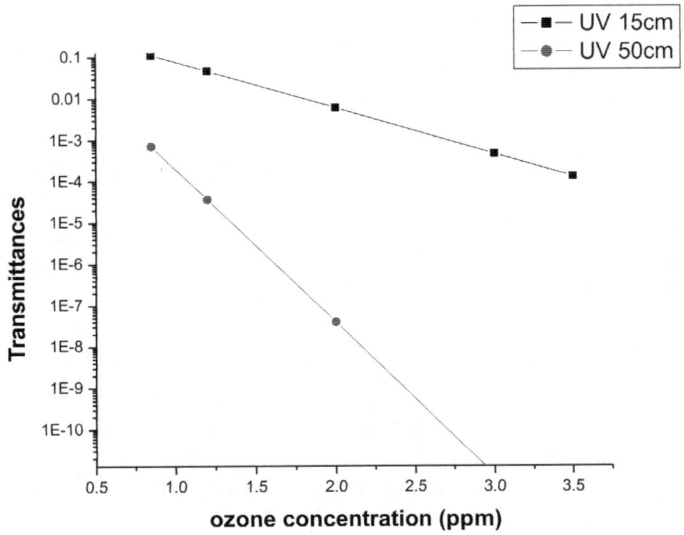

Figure (7) Transmittance of UV cell with different path length as a function of ozone concentration

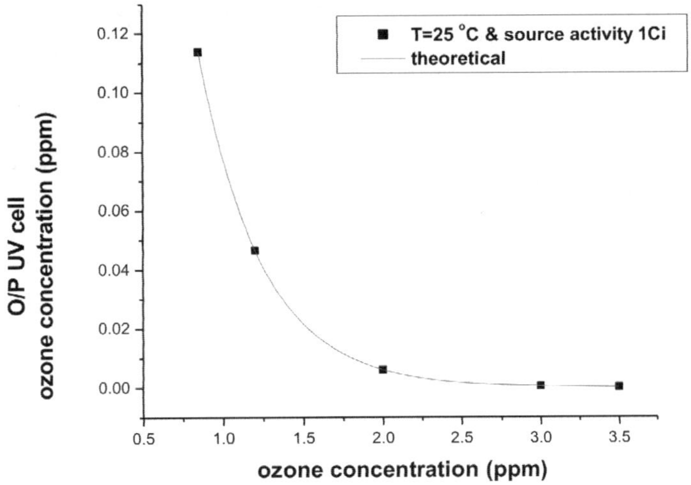

Figure (8) The relation between ozone concentration without UV and with UV
UV (254nm & 15 cm).

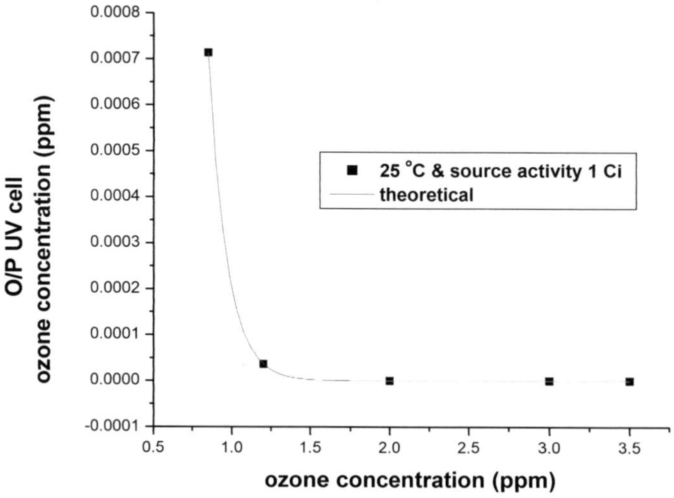

Figure (9) The relation between ozone concentration without UV and with UV UV (254 & 50 cm).

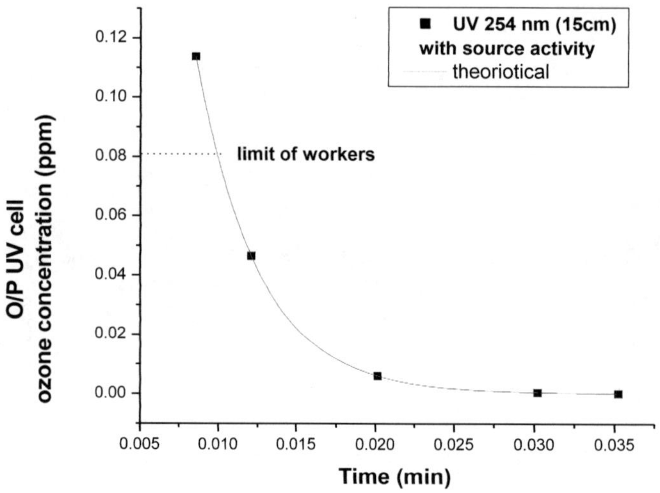

Figure (10) Ozone decay curve in exhaust pipe, with the exhaust system working and activity 1 Ci and V3 with flow rate 0 using UV (15cm) cell .

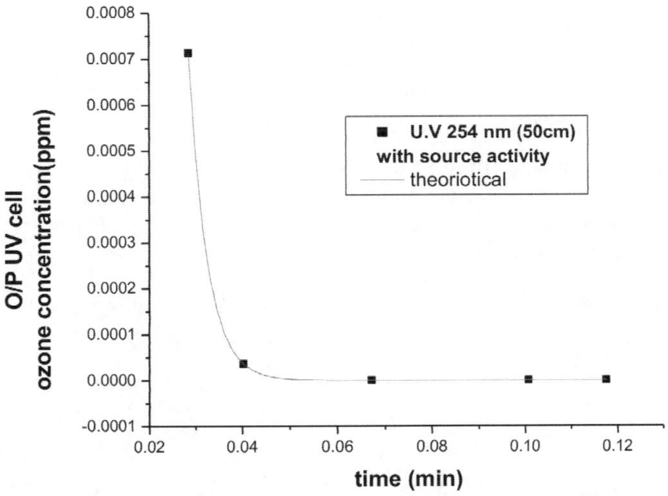

Figure (11) Ozone decay curve in exhaust pipe, with the exhaust system working and activity 1 Ci and V3 with flow rate 0 using UV (50cm) cell.

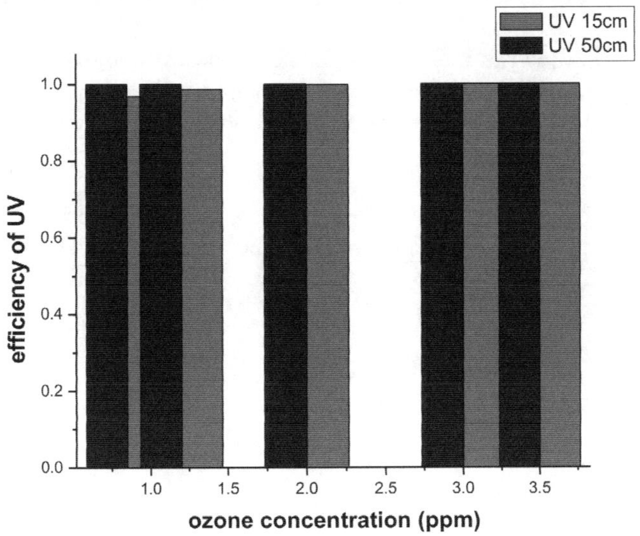

Figure (12) efficiency of using UV cell to reduce ozone concentration.

4. Conclusion

There is a positive correlation between the ozone concentration and UV absorption. A higher UV absorption resulted in a decreasing Transmittance slope, as UV is shown to absorb the ozone chemicals. Using a UV optical cell in the range of 15-50 cm illustrated decreasing ozone concentration levels. it is observed that 50 cm pathlength is the best wavelength yielding the highest efficiency of using a UV cell at approximately 99%. It is recommended to use 254 nm UV cells as a standby control system in CO60 industrial irradiators to reduce the decay time of the ozone inside the irradiation chamber. it is a cheap control system and useful when there is a defect in the exhaust system inside the radiation chamber.

5. References

[1] D. H. Uzueli1, 2013, "Ozone Concentration inside IPEN Industrial Irradiation Facility", Journal of Radiation Physics and Chemistry.

[2] EH40/2005," Workplace exposure limits: Containing the list of workplace exposure limits for use with the Control of Substances Hazardous to Health Regulations (as amended) Environmental Hygiene Guidance Note EH40 (Second edition) HSE Books 2011 ISBN 978 0 7176 6446 7 www.hse.gov.uk/pubns/books/eh40.htm

[3] Mar´c, M.; Tobiszewski, M.; Zabiegała, B.; de la Guardia, M.; Namie´snik, J.2015," Current air quality analytics and monitoring" A review. Anal. Chim. Acta , 853, 116–126. [Cross Ref] [PubMed]

[4] Marcus, T.C.E.; Ibrahim, M.H.; Ngajikin, N.H.; Azmi, A.I., 2015 "Optical path length and absorption cross section optimization for high sensitivity ozone concentration measurement". Sens. Actuators B Chem., 221, 570–575. [Cross Ref]

[5] Li, Y.; Dvoˇrák, M.; Nesterenko, P.N.; Nuchtavorn, N.; Macka, M.2018, "High power deep UV-LEDs for analytical optical instrumentation. Sens. Actuators B Chem. , 255, 1238–1243. [Cross Ref]

[6] Gordon, I.E.; Rothman, L.S.; Hill, C.; Kochanov, R.V.; Tan, Y.; Bernath, P.F.; Birk, M.; Boudon,V.; Campargue, A.;Chance, K.V.; et al. The HITRAN 2017," molecular spectroscopic database". J. Quant. Spectrosc. Radiat. Transf., 203, 3–69. [Cross Ref]

[7] Marcus, T.C.E.; Ibrahim, M.H.; Ngajikin, N.H.; Azmi, A.I. 2015,"Optical path length and absorption cross section optimization for high sensitivity ozone concentration measurement". Sens. Actuators B Chem., 221,570–575. [Cross Ref]

[8] Degner, M.; Ewald, H.; Lewis, E. 2011,"LED based spectroscopy-A low cost solution for high resolution concentration measurements e.g. for gas monitoring applications". In Proceedings of the Fifth International Conference on Sensing Technology, Palmerston North, New Zealand, 28 November–1 December ; pp. 145–150.

[9] Khan, S.; Newport, D.; Le Calvé, S.2019, "Development of a Toluene Detector Based on Deep UV Absorption Spectrophotometry Using Glass and Aluminum Capillary Tube Gas Cells with a LED Source. Micro machines", 10, 193. [Cross Ref] [PubMed]

Improved treatment of home-made glow discharge ion source

M M Abdel Rahman, F W Abdel Salam and B A Soliman
Accelerators & Ion Sources Department, Nuclear Research Center, Egyptian Atomic Energy Authority, P.O.Box: 13759, Cairo, Egypt.

E-mail: moustafa82003@yahoo.com

Abstract. For a better treatment, a new class of DC ion source (DCIS) was developed and adapted to provide low-power, highly-density plasma and ion-current applications. Such an ion source operates under the mode of abnormal glow discharge being of the self-extraction category, with specific features such as simple operation, stable discharge, and the ability to use for a variety of plasma, ion, and electron beam applications. A teflon disc with different diameters has been installed in between the two electrodes.This disc covered the cathode area, reducing the discharge area on the cathode surface for discharge confinement, and thus allowing for a higher output ion beam current. Pressures of 10^{-4} mbar, a 3 mm gap between the electrodes, and a 5 mm teflon diameter were observed to be the optimal operating conditions. At this ideal distance, hydrogen, nitrogen, and argon gases were used to investigate internal and external operational parameters. Finally, the Paschen curves were determined for hydrogen, air, as well as nitrogen gases.

Keywords: Pierce type ion source, Pachens curve and different gases, self-extraction type.

1. Introduction

Direct and ion implantation, thin layer sputter deposition, ion lithography, and etching are types of ion beam technologies. Such processes are using in material treatment for generating innovative features [1-5]. Mechanical, optical, magnetic, and electric properties of materials, can all be modified by ion beam treatment. As a result, ion beam research has recently evolved away from fundamental features of ion–solid interaction and toward surface engineering, particularly nanostructure technology, new tribiological, corrosion, and optical applications. Earlier references [6-8] provide detailed reviews of ion–solid surface interactions.

Glow discharge ion sources [9] have recently gotten a lot of press because of their several benefits, including the lack of expensive high-vacuum equipment, cheap total costs, simple systems, and ease of use. A glow discharge is a discharge in which the cathode emits electrons while being bombarded by particles and light quanta from the surrounding gas. Positive ion bombardment produces electrons at the cathode surface, which keep the discharge going.

The axial motion of charged particles transports current through such a glow discharge. The pattern of the electric field, or its axial component, can be used to understand the flow of current across the cathode zones. The electric field is found to be strong at the cathode, decreasing in intensity towards the negative light, and then remaining constant along the positive column after crossing a limited distance in the Faraday dark region. It only rises again at the anode. The availability of energetic particles for applications makes the cathode region of dc glow discharges important.

In the present work, we introduce miniaturized plasma and ion beam source that is stable with operating features for the discharge and output current. Finally, the Paschen curves are obtained at different gases of hydrogen, air and nitrogen.

Journal of Physics: Conference Series **2304** (2022) 012011 doi:10.1088/1742-6596/2304/1/012011

2. Theoretical investigation of gas flow at low pressures

The concepts for the gas pressure difference inside the ion source compared to outside the ion source have been theoretically examined. The flowing gas governs the source pressure and gas consumption by establishing pressure differential between the source and the high vacuum zone. On the other hand, at such low pressures as used in the home-made ion source, gas flow is considered in the cylindrical canal. The gas flow c is given by:

$$c = 3.64 * k' A \sqrt{T/M} \tag{1}$$

and ,

$$k' = \frac{1}{1 + 3/8\, L/a} \tag{2}$$

where,

T absolute temperature (k°)

M molecular weight of the gas

A is the cross section cm^2

L is the length of the canal and a radius.

The gas flow conductance for hydrogen at room temperature (T= 300 k°) and with M= 2.016 gm / mole, is given by:

$$c_{hyd} = \frac{444\pi a^2}{1 + 0.375 L/a} \tag{3}$$

where, a is the canal radius and L is length in millimeters.

The gas consumption Q (Q= cΔp), can be estimated as:

$$Q_{hyd} = \frac{6.62 a^2 p}{1 + 0.375 \frac{L}{a}} \tag{4}$$

and the maximum current that could be produced is:

$$I_{\max hyd} = 28.8 V^{3/2} \left(\frac{d}{L} \right)^2 \tag{5}$$

From eqs. 4, 5:

$$I_{\max hyd} = \frac{16.2 V^{3/2}}{\left(6.62 \frac{a^2 p}{Q} - 1 \right)^2} \tag{6}$$

Equation 6 estimates the highest beam current as a factor of acceleration voltage (between the electrodes), canal radius, gas consumption, and gas pressure.

Table 1: Values of pressures within ion sources and in vacuum chamber

C_T	Q	P (vacuum chamber) (mbar)	ΔP (mbar)	P (within the ion source) (mbar)
0.0026	0.0005	4 x 10^{-4}	2.75 x 10^{-1}	2.75 x 10^{-1}
0.0026	0.0008	6 x 10^{-4}	4.12 x 10^{-1}	4.12 x 10^{-1}
0.0026	0.0011	8 x 10^{-4}	5.49 x 10^{-1}	5.5 x 10^{-1}
0.0026	0.0014	1 x 10^{-3}	6.86 x 10^{-1}	6.87 x 10^{-1}

3. Experimental setup.

Figure 1 shows the schematic layout of a home-made glow discharge DCIS as well as its electrical circuit. That source, as first pointed out [10] and was needed to improve somewhat for modifications as increasing of their dimensions and addition of Teflon disk of various diameters to be able for the capabilities of various applications of plasma and ion beam.

This ion source is made up of a copper Pierce form anode with a tiny aperture that acts as a mechanical constriction, while the cathode is made of aluminium plate. The height of the anode is 34 mm. On the top section, it has a homogeneous cylinder with an inner diameter of 46 mm and a height of 24 mm, which is aligned to a cone shape with a 20 mm inner diameter on its outlet. The aluminum cathode's aperture is 2 mm diameter and 5 mm in height. A Bakelite envelope surrounds the electrode system. Faraday plate is used to monitor the output ion beam current and is located 5 cm from the cathode's ion exit opening. Between the electrodes, a Teflon disc of various sizes has been inserted. This disc covered the cathode surface, decreasing the discharge region on the cathode surface for discharge confinement, and therefore providing for a larger output ion beam current. Ionization is triggered when primary electrons hit gas particles due to an electric potential between electrodes. As a consequence, high beam current could be drawn longitudinally from the discharge zone without extraction system.

A 10 kV power supply is used to start the DC glow discharge between electrodes. The evacuation of ion source chamber is completed with diffusion pump.

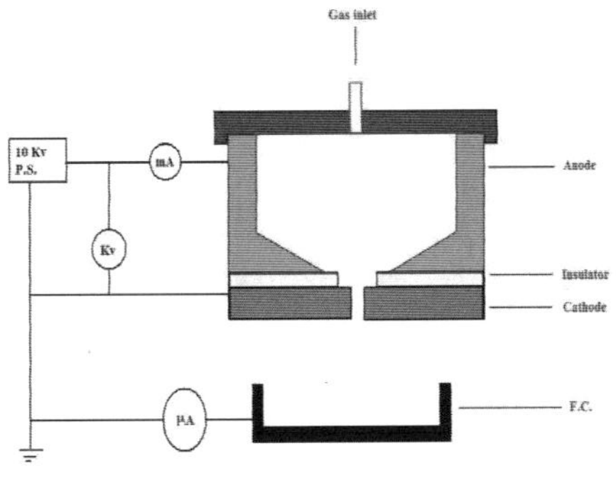

Fig. 1: Schematic of the ion source and its electrical circuit.

4. Experimental results and discussion

4.1 Optimization the spacing between ion source electrodes using hydrogen gas

A benefit from the idea of the Douplasmatron ion source construction (Filament, intermediate electrode of Pierce shape and plane anode), where this arc type ion source creates two regions of high density plasmas, as a result, a large output current can be achieved. Therefore, an Intermediate (Pierce shape) has been homely-demonstrated and built. As a result, without the need of an extraction equipment. In this work, the optimization of the distance between electrodes for the glow discharge ion source has been determined using hydrogen gas. Also, the ion source efficiency (I_d via I_b) is investigated under different experimental conditions.

Figure 2 shows the discharge properties of hydrogen gas at various gas pressures. In this figure, where the features of such a discharge are defined by abnormal glow [9], an increase in discharge voltage was noticed to be followed by an increase in discharge current.

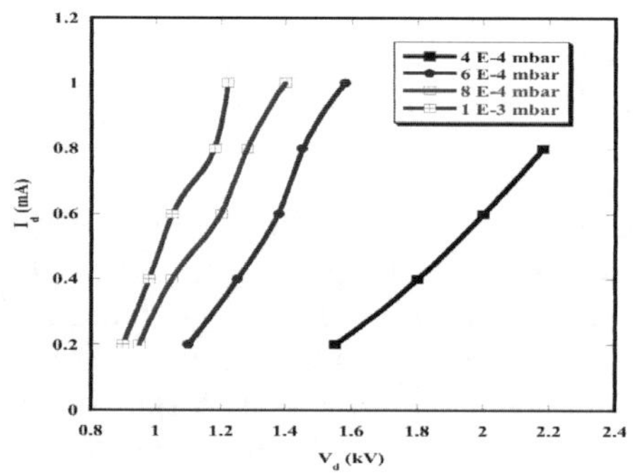

Fig 2: Discharge current vs discharge voltage at a distance of 3 mm between electrodes using hydrogen gas at various gas pressures.

The ion beam efficiency, (i.e. the relationship between ion current vis discharge current) at a 3 mm gap between electrodes at various gas pressures using hydrogen gas, is shown in Fig. 3. The output ion current increases as the discharge current increases, as can be seen in this figure. At a 3 mm gap spacing and a pressure of 4 x 10^{-4} mbar, the greatest values are attained. This ion source has a dominant confined behaviour due to a mechanical constriction at the lower end of the anode and a strong electric field strength in the tight gap.

At a gas pressure of $4x10^{-4}$ mbar , with hydrogen gas, Fig.4 shows the relationship between the output ion beam current and the discharge current, for various values of the gap distance between electrodes.

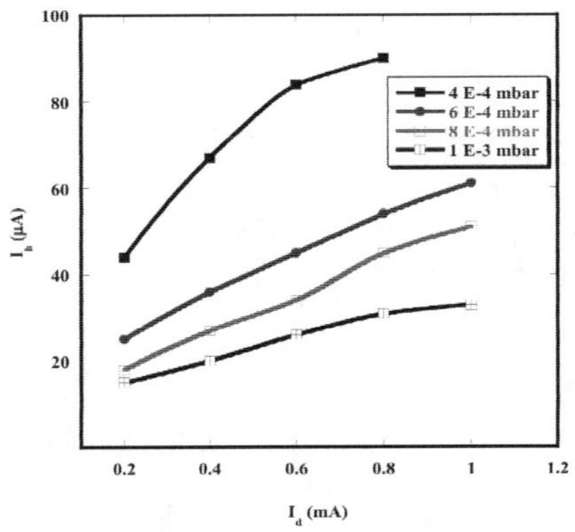

Fig. 3: Beam current as a function of the discharge current at various H_2 pressures and 3 mm gap between the electrodes.

the anode and strong electric field strength in the narrow gap, this ion source shows a preferred restricted behaviour.

The output ion current increases as the discharge current increases, as can be seen in this diagram. Also with a pressure of 4 x 10^{-4} mbar and a 3 mm electrode gap spacing, the maximum values are obtained.

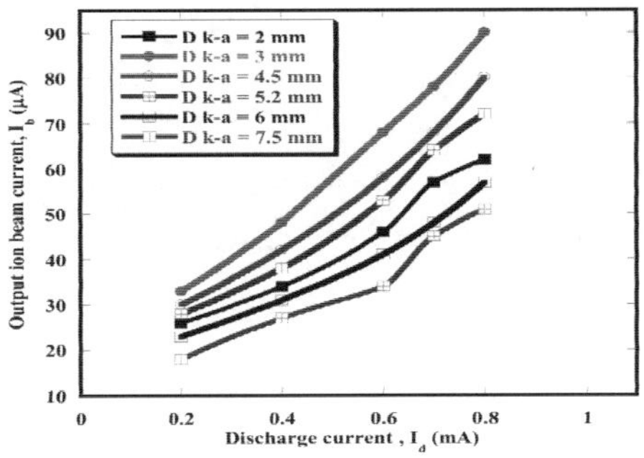

Fig. 4: The ion current vs. discharge current at different electrodes distances, using H_2 gas at pressure of 4x10^{-4} mbar.

Fig.5 shows the changes of the output ion current with gap distance between electrodes by using hydrogen gas at a discharge current of 0.6 mA and a pressure of 4 x 10^{-4} mbar. As seen in this plot, the increase in output ion current was accompanied by a decrease in gap distance. The ideal gap distance between electrodes in our experimental system was found to be 3 mm because the discharge is unstable when d is smaller than 2 mm.

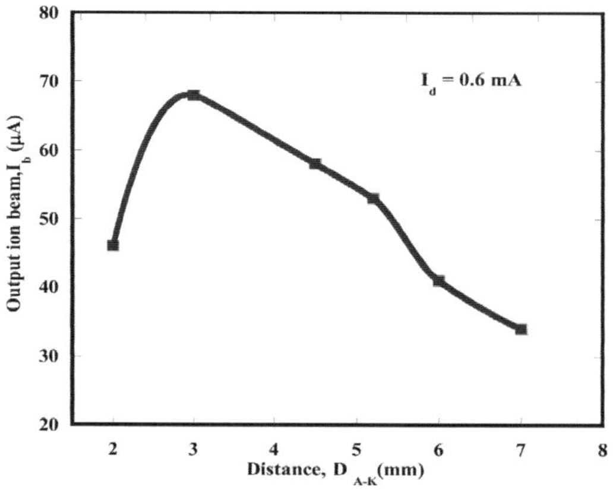

Fig. 5: The correlation between beam current and electrode gap at operating condition of $4x10^{-4}$ mbar and 0.6 mA.

At a pressure of $4x10^{-4}$ mbar, Figure 6 displays the output ion current vs. discharge current for various Teflon orifices. The ion current is 139 µA with a 5 mm orifice diameter and D_{k-a} = 3 mm At this orifice diameter, the highest ion current could be attained. It is clear that lowering the pressure enhanced both the output ion current and the discharge voltage, implying that reducing the working gas pressure improves the etching of any specimen exposed to output ion beam [11].

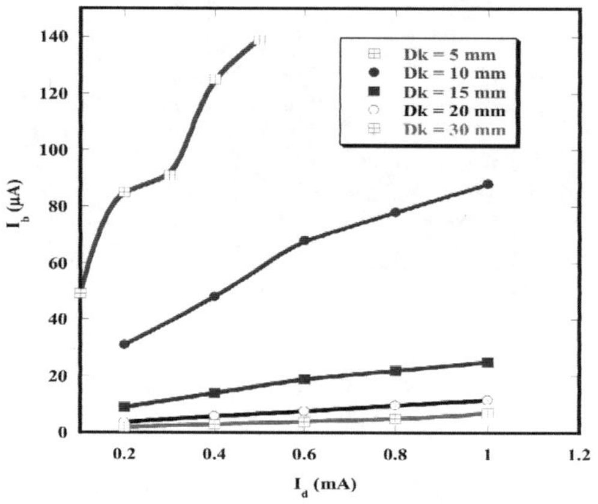

Fig. 6: Ion current as a function of discharge current at 3 mm distance between electrodes, at variable insulator diameters using H_2 gas at $4x10^{-4}$ mbar

The output ion beam current is shown against the diameters of the Teflon disc orifices in Figure 7. At a hydrogen gas pressure of 4×10^{-4} mbar and an orifice diameter of 5mm, a maximum output ion beam current may be attained.

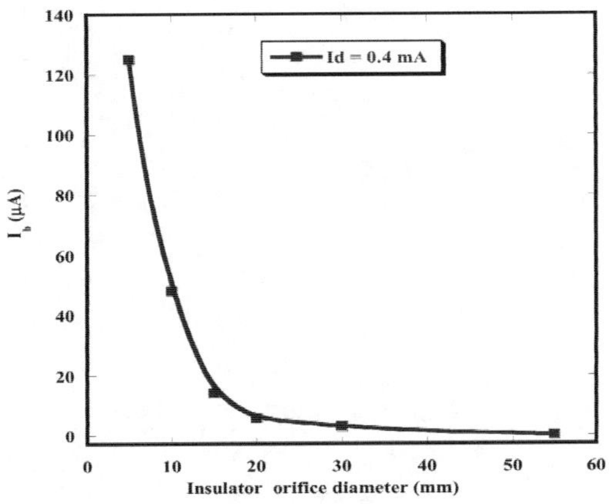

Fig. 7: The output ion current vs. the insulator Teflon diameters for H_2 at pressure of $4x10^{-4}$ mbar at discharge.

4.2. Operational characteristics using argon and nitrogen gases

Figure 8 presents the discharge current vs discharge voltage for electrodes distance = 3 mm using argon gas at various pressures. An ideal pressure of $6x10^{-4}$ mbar was determined, but high vacuum needs high voltage. Figure 9 shows the output ion current vs. discharge current for electrodes separated by 3 mm in argon gas at various gas pressures. The greatest ion current is achieved at a pressure of $4x10^{-4}$ mbar. Figure 10 shows the discharge current vs. discharge voltage for electrodes separated by 3 mm when nitrogen gas is used at various pressures. The optimal pressure was found to be $6x10^{-4}$ mbar, however high vacuum required high voltage. The same features are shown in Figure 11 employing nitrogen gas at a pressure of $6x10^{-4}$ mbar and electrode spacing of 3 mm.

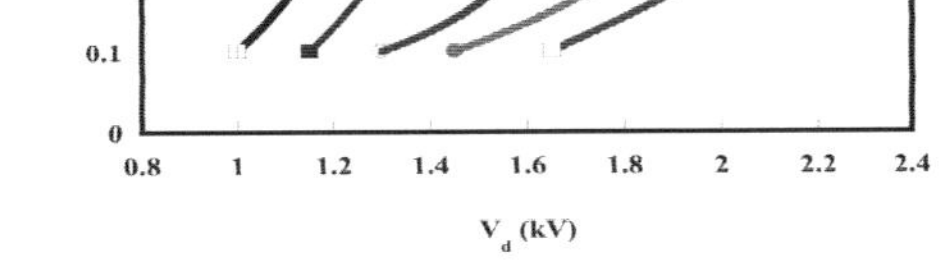

Fig. 8: Correlation between discharge current and the voltage at various H_2 pressures for electrodes gap of 3 mm using Ar gas.

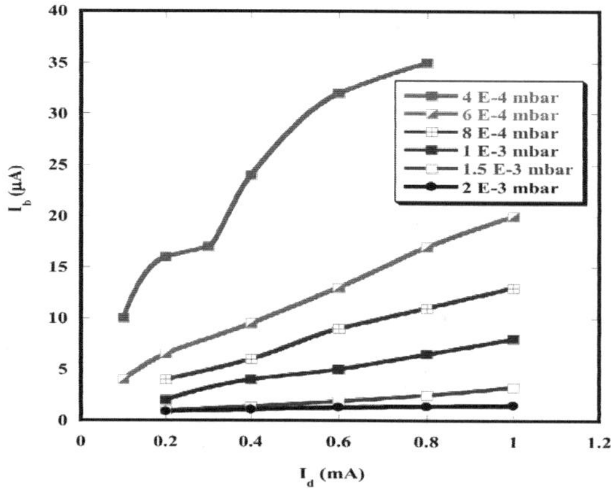

Fig. 9: Ion current vs. discharge current at various Ar pressures for 3 mm gap between electrodes.

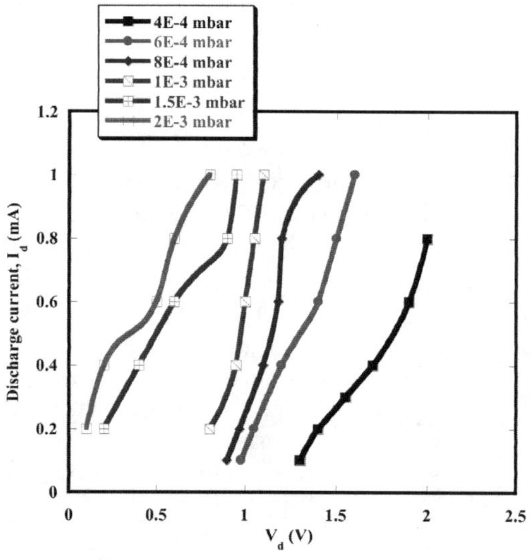

Fig. 10: Discharge current as a function of the discharge voltage at various N_2 pressures for 3 mm gap between electrodes.

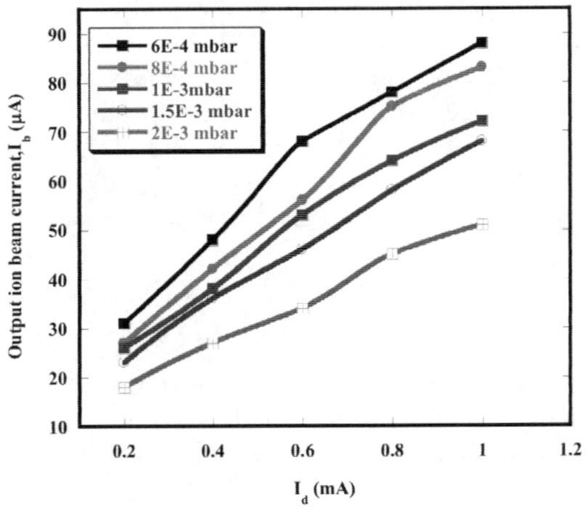

Fig. 11: Ion current vs. the discharge current at various N_2 pressures for 3 mm gap between electrodes.

4.3. Electrical breakdown and Paschen's curves

Glow discharges are commonly used for creating thin polymer and oxide layers, cleaning the surface of materials, and other applications, hence research into the conditions that cause glow discharge breakdown is of keen importance. As far as it is known, the breakdown curves of the glow discharge are described by Paschen's law. The effect of gas composition on the breakdown voltage is investigated in this paper. Paschen's curves were measured in the pressure range of 0.0001 to 0.00024 mbar with the inter-electrode distance set at 3 mm for varied H_2, N_2, and Air (Fig. 12). A common observation about these curves is that they all have virtually the same general behaviour and follow the standard curve. After reaching a minimal value, the breakdown voltage, V_b, falls with increasing Pd and then continues to climb after passing through a minimum value [12]. When Pd is increased, V_b rapidly decreases on the left-hand side of the minimum Paschen's curves, which can be attributed to an increase in collision frequency, which is related to an increase in the number of electron-neutral atom or molecule collisions. V_b increases slowly as Pd increases on the right-hand side of the minimum, showing that the ionisation cross-section grows. As a result, electrons require more energy (increasing V_b) to breakdown the discharge gap[13].

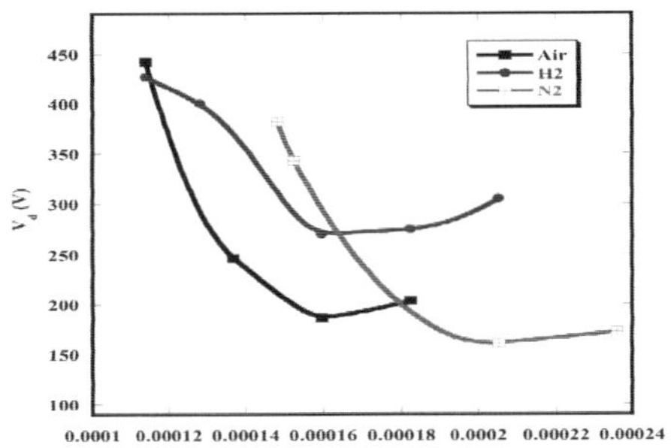

Fig. 12: Electrical breakdown and Paschen's curves for H_2, N_2 and Air gases.

Conclusion

A home-made glow discharge ion source with a relatively large output ion current has been designed and created. This ion source is small in size and has simple and long-term steady operation in a wide range using different gases; there are no magnetic fields or filaments in this source. It's also been demonstrated that the present ion source can produce a various types of ion beams using Hydrogen, Nitrogen, Argon and Air.Teflon diameter of 5 mm (the diameter that covers the cathode surface) was found to be the best diameter for the ion source, giving the highest ion beam currents. Electrical breakdown and Paschen's curves were measured for H_2, N_2, and Air gases, with air and hydrogen gases having a higher voltage than nitrogen gas. This ion source is suitable for applications like as etching, sputtering, and micromachining.

Acknowledgements

For its financial support of this endeavor, the Egyptian Atomic Energy Authority deserves special mention. The authors acknowledge Mr. I. Abdelbaki and Mr. M. Madboli with expertly manufacturing parts for the direct current ion source (DCIS).

References

[1] Wolf B.H 1995 Handbook of Ion Sources, CRC press, Boca Raton, New York.
[2] Clark D.J 1983 Proc. of the International Ion Engineering Congress, Kyoto, Japan, 12–16 Sept.
[3] Brown I.G 2004 in The Physics and Technology of Ion Sources, I. G. Brown (Ed.), John Wiley, New York.
[4] Valyi L 1977 Atom and Ion Sources, John Wiley & Sons, New York.
[5] Angert N 1994 Ion Sources, Proc. CERN Fifth General Accelerator Physics Course, Jvyaskla, September, CERN 94–01, 619.
[6] Oks E 2006 ‘‘Plasma cathode electron sources, physics, technology, applications". Weinheim, Germany: Wiley VCH Verlag GmbH & Co. KGaA.
[7] Miljevic V. I 1990 Rev. Sci. Instrum. **61**, 312-314.
[8] Anders A and Kühn M 1997 7th International Conference on Ion Sources Taormina, Italy, September 7-13.
[9] Marcus R.K and Broekaert JAC. 2003 ‘‘Glow discharge plasmas in analytical spectroscopy’’. John Wiley & Sons.
[10] Abdel Rahman M.M, Helal A. G, Moustafa O.A and Abdel Salam F.W 2008 Journal of Nuclear and Radiation Physics **3** 1
[11] Wilson R.G and Brewer G.R 1973 Ion Beam with Applications to Ion Implantation, John Willy and Sons, New York.
[12] Lisovskiy V. A, Yakovin S. D and Yegorenkov V. D 2000 J. Phys. D: Appl. Phys **33** 2722
[13] Von Engel A 1983 "Electric Plasmas: their nature and uses", Taylor & Francis, London.

Utilization of INAA Technique for Studying the Negative Impact of Human Activities on Coral Reefs in the Quseir harbour area

Mohamed Y.M. Mohsen[1*], Atef El-Taher[2], Mohamed Shabib[3], Nader M. A. Mohamed[4], H. A. Ashry[3], Mohamed A.E. Abdel-Rahman[1]

[1*]Nuclear Engineering Department, Military Technical College, Kobry El-kobbah, Cairo, Egypt
[2]Physics Department, Faculty of Science, Al Azhar University, Assiut 71524, Egypt.
[3]Department of Radiation Physics, National Center for Radiation Research and Technology (NCRRT), Egyptian Atomic Energy Authority (AEA). 3Ahmed El-Zomor St. Nasr City El-Zohoor District, P. O. 11787. Cairo, Egypt.
[4]Reactors Department, Egyptian Atomic Energy Authority, Egypt, P.O. Box 13759, Cairo, Egypt.

Email: moh39333@mtc.edu.eg

Abstract. Coral reef samples collected from the Egyptian Red Sea coast's Quseir harbour were analysed using the INAA technique for qualitative and quantitative analysis. The concentration of heavy metals, rare earth elements, and elemental contents were measured in coral reefs. The samples were assembled with their standards and irradiated in Egypt's Second Research Reactor at the same time (ETRR-2). The gamma-ray spectra of an HPGe detector with a 100 percent relative efficiency were studied. The elements determined are: Al, Ti, Na, K, Mg, V, Cl, Ca, Mn, Cr, As, Co, Zn, Fe, Br, Tb, Zr, Hf, Sb, Ba, Sc, Nd, La, Sm, Ce, Eu, Yb, Ta, U, and Th. The present work is worried about the use of INAA to attain an accurate knowledge about the elemental content in coral reef samples in the Quseir harbour area and revealing potential environmental indicators for heavy metals contamination as well as pollution hot spot throughout this area to assess the public health for workers, tourists, and the surrounding population.

Keywords: Coral Reefs Samples- Neutron Activation Analysis – Quseir Harbour area.

1. Introduction

The Red Sea is an Indian Ocean seawater entrance located between Asia and Africa. The Sinai Peninsula, the Gulf of Suez, and the Gulf of Aqaba are all to its north. The Red Sea also includes large shallow shelves that are known for their corals and marine life. Over 1,000 invertebrate species and 200 different types of hard and coral can be found in the sea.

Coral reefs are one of the planet's oldest ecosystems. They are enormous limestone formations that provide habitat for almost a fifth of all aquatic species. Carbonates make up the skeletons of hard corals. In general, herma-typic corals or reef-building corals are referred to as corals. They come in a variety of shapes and sizes, including gigantic, branching, encrusting, solitary, and so on. Carbonate is primarily derived from corals in shallow marine sediments. Because coral reefs are popular tourist destinations, their environmental status is crucial. Sediment run-off, water pollution, collecting and dredging, explosive fishing, reckless recreation, and global warming are all significant hazards to coral reefs. [1-2].

Different procedures for assessing the trace, minor, and major constituents of these environmental marine samples, which are considered composite samples, could be used. NAA's key advantages include high-precision interferences and relative independence from matrix effects. Because nuclear reactions and decay processes are almost unaffected by the chemical and physical structures of the substance during and after irradiation, the matrix composition has a minor impact on the samples after activation. [3].

The aim of the present work is to use INAA to obtain precise information about the elemental content in coral reef from Quseir harbour and classify potential environmental indicators for heavy metal contamination that influence workers, tourists, and the surrounding population.

2. Experimental work

2.1. Samples Preparation and Irradiation

Eleven samples of living coral reef skeletons were collected from Quseir harbour including (ex: Stylophora subseriata, Acropora forskalli, Porites solida, Favia albidus). Figure1 showing distribution of bottom facies of the Quseir harbour. Figure **Error! No text of specified style in document.** shows different types of photos of Quseir Harbour.

All coral reef samples were collected by SCUBA diving into the interior of the sea. The collected samples were mashed, homogenized, and sieved. From about 150 mg to about 250 mg of the sample materials were packed in high purity polyethylene capsules for short and long irradiation. For short irradiation, a computerized pneumatic rabbit system was used to send polyethylene capsules containing one sample into an irradiation position. Samples and standard reference materials (Soil powder (IAEA-Soil-7), Whey powder (IAEA-155), Hay powder (IAEA-V-10), and NIST standard were individually irradiated at a suitable thermal neutron flux $4\times10^{11}cm^{-2}s^{-1}$ for 3min in a pneumatic system at Egypt Second Research Reactor (ETRR-2). The samples were measured two times, after 5mins and after 5hrs while the counting times were 5 mins and 15 mins respectively. During the counting of short-lived radionuclides, the dead time often decreases considerably. For long irradiation, the samples were packed in two aluminium cans (two groups). Each can contain 9 samples in high purity polyethylene capsule, together with the same certified reference materials like in short irradiation and an empty polyethylene bag (the "blank") was sealed. The samples and standards are irradiated at a thermal neutron flux $1.5x10^{14}cm^{-2}s^{-1}$ for 30mins. The samples were counted two times, after 7d and after 50d while the counting times were 30mins and 1hrs, respectively. Table1 shows the irradiation cycles. Table 2 gives the nuclear properties of the determining elements in short and long irradiation time.

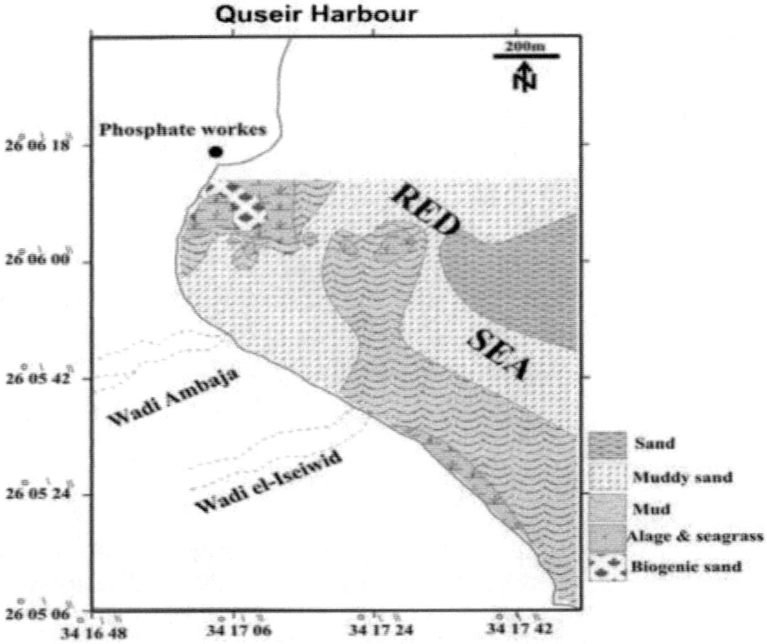

Figure 1 Under most facies of the Quseir harbor.

Table1 Irradiation cycles,

Cycles	run	Irradiation Time	Counting Time	Average Cooling Time
Short	1st	3min	5min	5min
	2nd	3min	15min	5hour
Long	1st	30min	30min	7day
	2nd	30min	1hour	50day

Table 2 Nuclear properties of the determining elements in short and long irradiation time.

Element	Product	Energy Line (keV)	$T_{1/2}$	Detection Limits (mg/kg)	Type of Irradiation
Al	Al-28	1779	2.25m	13	Short
Ca	Ca-49	3083	8.72m	957	Short
Cl	Cl-38	2167	37.2m	283	Short
Mg	Mg-27	1014	9.5m	354	Short
Na	Na-24	1369	15h	60	Short
V	V-52	1434	3.75m	0.5	Short
K	K-42	1524	12.7h	108	Short
Mn	Mn-56	846	2.6h	0.02	Short
Ti	Ti-51	320	5.8m	0.4	Short
As	As-76	559	1.08d	0.3	Long
Ba	Ba-131	496	11.8d	33	Long
Br	Br-82	776.6	35.3 h	0.2	Long
Ce	Ce-141	145	32.5d	0.2	Long
Co	Co-60	1332	5.27 y	0.03	Long
Cr	Cr-51	320	27.7d	0.9	Long

Eu	Eu-152	1408	13.5 y	0.01	Long
Fe	Fe-59	1291.6	44.5 d	62	Long
Hf	Hf-181	482	42.4d	0.4	Long
La	La-140	1596	1.68 d	0.12	Long
Sc	Sc-46	1120.5	83.81d	0.0008	Long
Sm	Sm-153	103	46.28h	0.01	Long
Ta	Ta-182	1221	114 d	0.001	Long
Tb	Tb-160	879	72.3d	0.04	Long
Th	Pa-133	312	26.9d	0.14	Long
U	Np-239	277	2.36d	0.4	Long
Zn	Zn-65	1115	243.9d	0.08	Long
Zr	Zr-95	756	64d	11	Long

2.2. Measuring system

The n-type hyper pure germanium (HPGe) detector with relative efficiency of 100 percent was used. A computer-based PC A-MR 8192 multi-channel analyser is included. The detector was housed in a 100 mm thick cylindrical lead shield. This detector has a resolution of 2.1 keV at 1.33 MeV and a peak to Compton ratio of 64, respectively. -4000V dc is the operating bias voltage. A blank container was used to take the backdrop measurement (i.e., empty). The net count for each radionuclide of interest was calculated by subtracting the background counts from the sample counts. The obtained γ-ray spectra were analysed using gamma vision computer software. The detector's energy and efficiency calibrations were carried out with standard point sources of 137Cs, 133Ba, 60Co, and 152Eu; all of them were produced by Isotope Products Laboratories [4].

2.3. Calculation of Concentration

It is usually more convenient to use a relative (comparative) method whereby the unknown sample is irradiated together with a standard sample containing a known concentration of the element (s) of interest. The standard sample is measured under the same conditions as the standard sample (sample-to-detector distance, sample size, and composition) [5].

From the comparison of the count rate Ap in the two measured spectra, ρsam and ρref (mg/kg) are the concentration of the element in unknown samples and the concentration of standard materials, respectively. The concentration of the element of interest can be determined by:

$$\rho_{sam} = \frac{\left(\frac{A_p}{SDCW}\right) \text{sample}}{\left(\frac{A_p}{SDCw}\right) \text{standard}} \times \rho_{Ref} \quad (1)$$

Where w and W are the masses of the standard material and the mass of the unknown sample, respectively. Decay factor (D), Saturation factor (S), and counting factor (C) are represented by the following equations:

$$S = \left(1 - e^{-\lambda t_{irr}}\right), \qquad D = e^{-\lambda t_d}, \qquad C = \frac{1 - e^{-\lambda t_c}}{\lambda t_c} \quad (2)$$

Journal of Physics: Conference Series **2304** (2022) 012012 doi:10.1088/1742-6596/2304/1/012012

In this procedure, many parameters are eliminated during the calculation of the concentration and other parameters are well known. The disadvantage of this method is that needs a standard (s) that contains all the elements of interest.

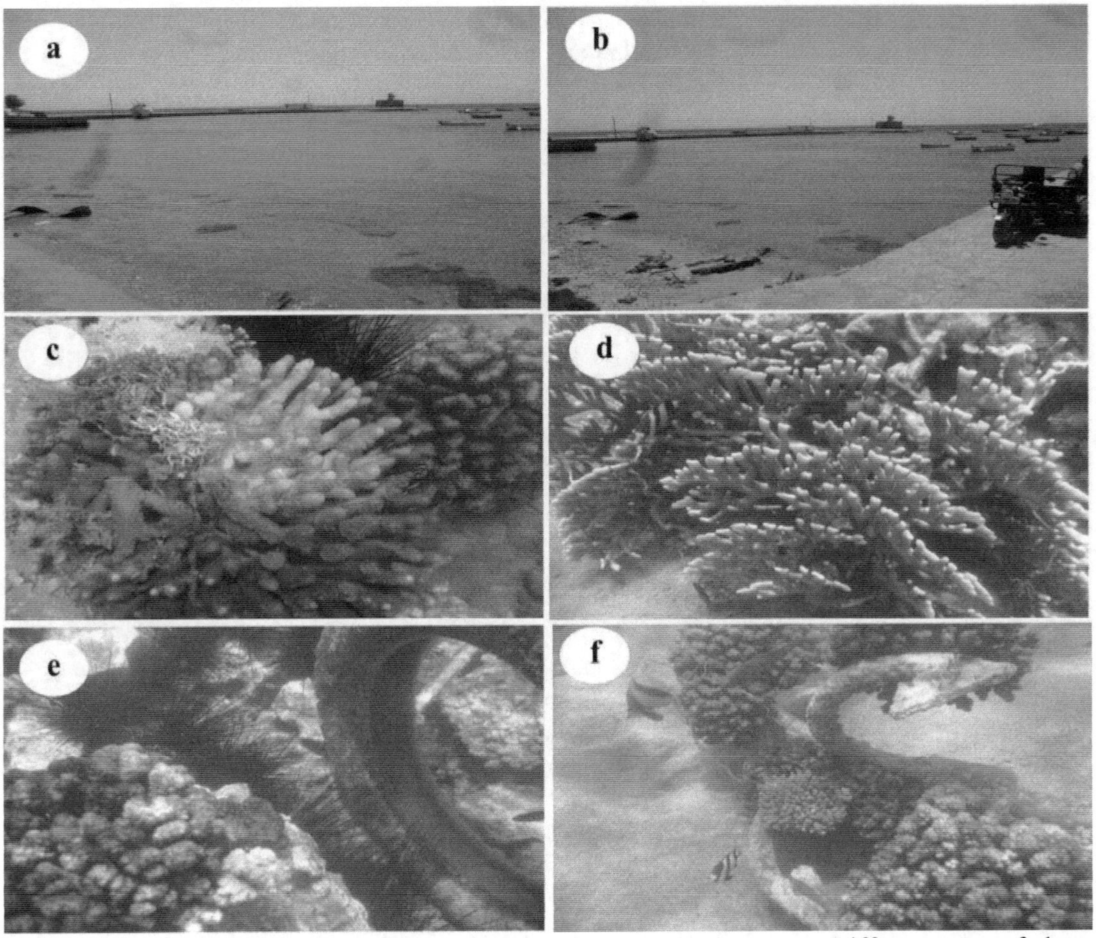

Figure Error! No text of specified style in document. Different types of photos of Quseir Harbour

3. Results and discussion

Table 3 gives the complete data of elemental analysis that existed in the Quseir harbour area. The average concentrations values of all elements are expressed in units of mg/kg. Table 4 gives the maximum, minimum, and average concentration for every detected element in this area. Figures 3 and 4 as examples of the typical gamma-ray spectrum of coral reef samples in the Quseir harbour area for short and long irradiation, respectively.

Journal of Physics: Conference Series **2304** (2022) 012012 doi:10.1088/1742-6596/2304/1/012012

Table 3 The maximum, minimum, and average concentration for every detected element in Quseir harbour area.

Element	Min.	Max.	Average
Al	96±5	3487±74	719±24
As	0.73±0.02	2.27±0.07	1.3±0.03
Ba	92±22	141±39	111±61
Br	13.2±3	65.26±12	30±6
Ca	259621 ± 8700	384395 ± 2308	323991.2 ±1204
Ce	0.99±0.1	3.82±0.2	1.9±0.18
Cl	3462±94	12156±280	6676±151
Co	0.12±0.04	0.92±0.05	0.3±0.04
Cr	2.17±0.2	8.38±0.8	3.6±0.3
Eu	0.01±0.005	0.13±0.01	0.04±0.01
Fe	61±3	2007±22	450±9.5
Hf	0.46±0.01	3.65±0.14	1.9±0.06
K	446±155	4024±357	1061±262
La	0.14±0.02	2.06±0.03	0.6±0.009
Mg	1323±66	18844±456	6075±144
Mn	1±0.08	33±0.5	7.5±1.67
Na	6062±47	10388±215	8245±163
Nd	13.39±1.4	16±1.6	12.5±1.3
Sb	0.2±0.08	0.2±0.08	0.2±0.08
Sc	0.02±0.002	0.8±0.05	0.2±0.01
Sm	0.11±0.01	0.55±0.05	0.27±0.002
Ta	0.03±0.02	0.14±0.02	0.1±0.02
Tb	0.05±0.01	0.05±0.01	0.05±0.01
Th	0.15±0.009	0.52±0.04	0.26±0.02
Ti	130±24	416.5±69	207±33
U	1.5±0.16	4±0.3	2.8±0.26
V	0.82±0.03	8.1±0.47	2.5±0.15
Yb	0.41 ± 0.03	0.41±0.03	0.41±0.03
Zn	5.94 ± 0.03	119±0.5	23±0.13
Zr	42 ± 3	164±12	92±7

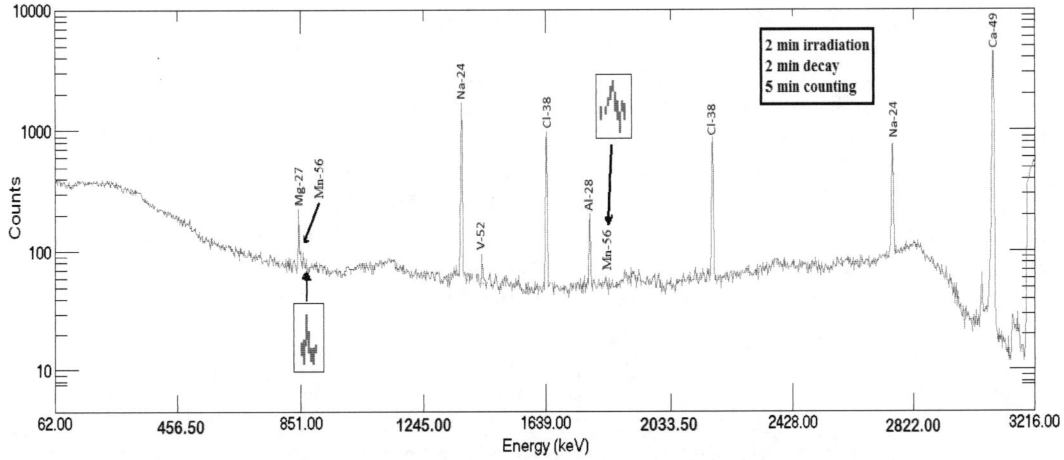

Figure 3 Gamma-ray spectrum of coral reef samples in Quseir harbour area after 2min irradiation, 5 min decay and 5 min counting time.

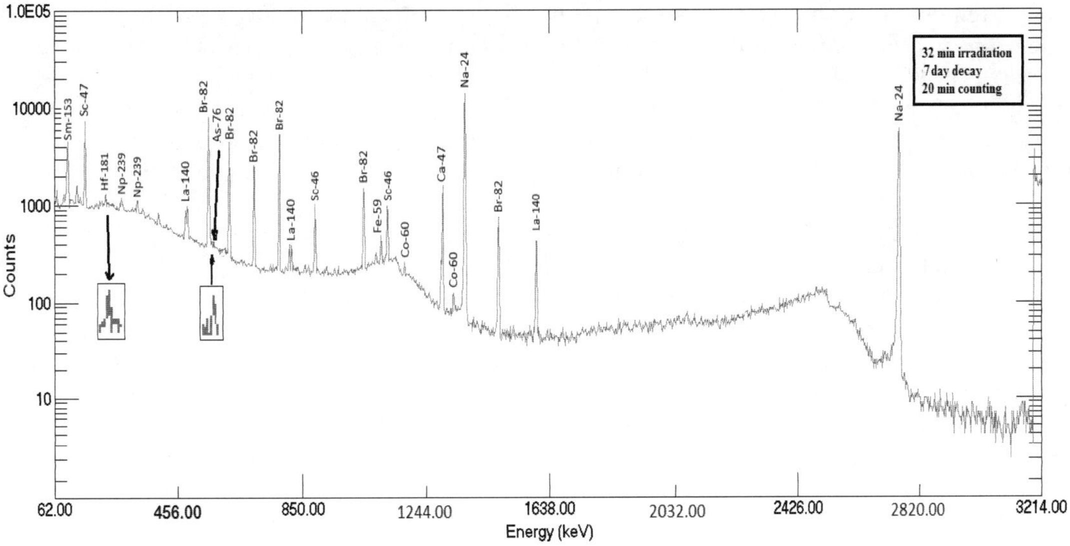

Figure 4 Gamma-ray spectrum of coral reef samples in Quseir harbour area after 32 min irradiation, 7-day decay and 20 min counting time.

Through a review of Table 3, it's noted that the average Calcium concentration equals (323991 ± 1204 mg/kg). This is attributed to the fact that Calcium is the main component in coral reefs. The average concentration of Na and Cl are (8245±163 and 6676±151 mg/kg), respectively, refer to the presence of sodium chloride in the seawater. The concentration of Mg reaches a maximum of 18844±456 mg/kg with a mean value of 6075±144 mg/kg. Calcium and Magnesium (Mg) are extremely similar chemically. In fact, they get in the way of each other in a diversity of situations, and that is part of the reason why these ions merit interest by aquarists. The excess in the concentration of Mg attributed to the substitution process with Ca for similarity in chemical properties [6].

Heavy metals come into the environment in two ways; naturally and anthropogenically as; municipal wastewater, manufacturing industries, and agricultural activities, etc. [7]. Various anthropogenic and natural impacts on the marine environment of the Red Sea including the contamination by oil slicks, wastewater release, effluents of desalination plants, building activities along the seashore, and landfilling and digging operations [8].

The rising of iron in both water and sediments might reflect the bio-accumulation in marine organisms as; coral reefs, bivalves, gastropods, and others [9]. Manganese concentration in coral reefs is an indicator of waste inputs [10]. This result indicates the high diversity and wide range in concentration of Mn in this area. Zinc is a fundamental micronutrient of coral reefs and also commonly used for cell repair and protein synthesis [11-12]. Zinc measured in coral reefs largely have anthropogenic or polluted sources and is enhanced in the coral's skeleton compared with the surrounding seawater [13]. The high concentrations of Zinc in the coral skeletons refer to the municipal wastes, industrial activities, and in the case of the coastal sediments [14-15]. The chronic toxicity of cobalt is that the chemical similarity with calcium and magnesium, which allow cobalt to accumulate in the skeletal matrix of many marine organisms like corals, bivalves, and crustacean skeletons [16].

Of the diverse sources of the Arsenic (As) in the environment, drinking water possibly poses the highest threat to human life [17]. The distribution and levels of Arsenic pollution in coral reef samples were assessed. The mean concentration of As in the Quseir harbour area is 1.3±0.03 mg/kg. The concentration of as was under the detection limit. The average concentrations of As in coral reef samples is still very limited by comparing previous studies [18-19].

Aluminium (Al) was long considered virtually not harmful to marine organisms, in large part because of the very low concentrations of aluminium naturally occurring in marine environments. It is now clear, however, that human activities are contributing to high concentrations of aluminium occurring in industrialized harbours and coastal waters, which have the potential for adverse impacts on marine organisms [20]. The mean concentration of aluminium in the Quseir harbour area is (719±24 mg/kg). The detected amount of Al may be referred to as the activity in the harbour and shipment wastes.

Determining the presence of major and trace elements, particularly rare earth elements (REEs), is crucial for a better understanding of the geochemistry of marine sediments [21] that may be deposited on coral reefs from an ecological and geochemical standpoint. Ce, Eu, La, Nd, Sm, and Sc had average concentrations of (1.9±0.18), (0.04±0.01), (0.6±0.009), (12.5±1.3), (0.27±0.002), (0.2±0.01), mg/kg, respectively. Tb and Yb concentrations were found to be 0.05±0.01 and 0.41±0.03 respectively. The assessment of the content of Th and U is essential from a radiological standpoint, since it indicates the radioactive influence on coastal marine sediments [22]. In the Quseir harbour area, the average concentrations of U and Th were 2.8±0.26 mg/kg and 0.26±0.02 mg/kg, respectively.

4. Conclusion

Using the NAA technique is considered a real success in the scope of monitoring the degree of environmental radioactive pollution. That was by determining the concentrations of heavy elements, and REEs elements with high accuracy and high precision. The measured concentrations of some heavy metals, rare earth elements, and some toxic elements were all within the acceptable limits with comparing previous studies. This means that no human activity is a hazard indicator for the surrounding environment. This work provides baseline data for element concentrations for future multi-element analysis of coral reef samples.

Journal of Physics: Conference Series **2304** (2022) 012012 doi:10.1088/1742-6596/2304/1/012012

References

[1] H. A. Madkour and M. A. Dar, "The anthropogenic effluents of the human activities on the Red Sea coast at Hurghada harbour (case study)," 2007.

[2] W. M. Badawy, A. El-Taher, M. V Frontasyeva, H. A. Madkour, and A. E. M. Khater, "Assessment of anthropogenic and geogenic impacts on marine sediments along the coastal areas of Egyptian Red Sea," Appl. Radiat. Isot., vol. 140, pp. 314–326, 2018.

[3] A. Abdulaziz and A. El-Taher, "Elemental analysis of basalt by instrumental neutron activation analysis and inductively coupled plasma mass spectrometer.," J. Environ. Sci. Technol., vol. 9, no. 4, pp. 335–339, 2016.

[4] Mohamed Shabib., Atef El-Taher., Nader M. A. Mohamed., Hashem A. Madkour., H. A. Ashry., 2021 Assessment of radioactivity concentration of natural radionuclides and radiological hazard indices in coral reefs in the Egyptian Red Sea. Journal of Radioanalytical and Nuclear Chemistry. 329:1199–1212

[5] C. Vandecasteele and C. B. Block, Modern methods for trace element determination. John Wiley & Sons, 1997.

[6] R. Holmes-Farley, "Aquarium chemistry: magnesium in reef aquaria," Adv. Aquar. Online Artic. http//www. Adv. com/2003/10/chemistry, 2003.

[7] A. M. Mansour, M. S. Askalany, H. A. Madkour, and B. B. Assran, "Assessment and comparison of heavy-metal concentrations in marine sediments in view of tourism activities in Hurghada area, northern Red Sea, Egypt," Egypt. J. Aquat. Res., vol. 39, no. 2, pp. 91–103, 2013.

[8] A. El-Sikaily, A. Khaled, and A. El Nemr, "Heavy metals monitoring using bivalves from Mediterranean Sea and Red Sea," Environ. Monit. Assess., vol. 98, no. 1, pp. 41–58, 2004.

[9] A. M. Dar, "Geological basis to study the environmental defect in the marine ecosystem as a result of tourist activities in Hurghada area and surroundings, Red Sea, Egypt," Ismailia, Egypt Suez Canal Univ., 2002.

[10] L. J. Linn, M. L. Delaney, and E. R. M. Druffel, "Trace metals in contemporary and seventeenth-century Galapagos coral: Records of seasonal and annual variations," Geochim. Cosmochim. Acta, vol. 54, no. 2, pp. 387–394, 1990.

[11] B. E. Brown and S. Howard, "Responses of coelenterates to trace metals: a field and laboratory evaluation," in Proc Fifth Int Coral Reef Congr, 1985, vol. 6, pp. 465–470.

[12] A. A. M. Ali, M. A. Hamed, and H. Abd El-Azim, "Heavy metals distribution in the coral reef ecosystems of the Northern Red Sea," Helgol. Mar. Res., vol. 65, no. 1, pp. 67–80, 2011.

[13] S. J. Fallon, J. C. White, and M. T. McCulloch, "Porites corals as recorders of mining and environmental impacts: Misima Island, Papua New Guinea," Geochim. Cosmochim. Acta, vol. 66, no. 1, pp. 45–62, 2002.

[14] M. Inoue, A. Suzuki, M. Nohara, H. Kan, A. Edward, and H. Kawahata, "Coral skeletal tin and copper concentrations at Pohnpei, Micronesia: possible index for marine pollution by toxic anti-biofouling paints," Environ. Pollut., vol. 129, no. 3, pp. 399–407, 2004.

[15] R. B. Clark, "Marine Pollution, Clarendon." Oxford, 1992.

[16] R. P. Smith, A primer of environmental toxicology. Lea & Febiger, 1992.

[17] S. Water and W. H. Organization, "Guidelines for drinking-water quality [electronic resource]: incorporating first addendum. Vol. 1, Recommendations," 2006.

[18] W. Maher and E. Butler, “Arsenic in the marine environment,” Appl. Organomet. Chem., vol. 2, no. 3, pp. 191–214, 1988.

[19] S. Khokiattiwong, N. Kornkanitnan, W. Goessler, S. Kokarnig, and K. A. Francesconi, “Arsenic compounds in tropical marine ecosystems: similarities between mangrove forest and coral reef,” Environ. Chem., vol. 6, no. 3, pp. 226–234, 2009.

[20] M. Gillmore, “TOXICITY OF ALUMINIUM IN SEAWATER: DIATOM SENSITIVITY,” 2014.

[21] A. El-Taher, W. M. Badawy, A. E. M. Khater, and H. A. Madkour, “Distribution patterns of natural radionuclides and rare earth elements in marine sediments from the Red Sea, Egypt,” Appl. Radiat. Isot., vol. 151, pp. 171–181, 2019.

[22] R. Khan et al., “Assessment of rare earth elements, Th and U profile of a site for a potential coal based power plant by instrumental neutron activation analysis,” Radiochim. Acta, vol. 106, no. 6, 515–524, 2018.

Effective Hybridization of Biogeography Based Optimization and Simulated Annealing

E G Zahran[1], A A Arafa[1], H I Saleh[1], M I Dessouky[2]

1 Department of Radiation Engineering, National Center for Radiation Research and Technology (NCRRT), Egyptian Atomic Energy Authority, Cairo, Egypt

2 Department of Electrical Communications, Faculty of Electronic Engineering Menoufia University Menouf, Egypt

E-mail: h_i_saleh@hotmail.com

Abstract. Since the optimization process constitutes a great step in solving complex real world problems, the development of novel optimization algorithms is one of the growing interest topics that attracted researchers in the recent decades. This paper presents the hybridization of bio-inspired Biogeography Based Optimization (BBO) algorithm and physics-inspired Simulated Annealing (SA) algorithm, into a new variant called BBO-SA. The proposed algorithm uses the concepts of the SA to enhance the diversity of BBO solutions which in turn improves the obtained solution. For validating the performance of BBO-SA, it is compared to that of BBO algorithm in solving a set of thirteen complex benchmark functions. Validation results prove the superior performance of the proposed BBO-SA algorithm over the BBO algorithm in solving complex function in terms of escaping from local optima and reaching near optimal solution in lower execution times. Besides, the proposed algorithm is applied to solve a very challenging problem denoted as the RFID Reader Deployment Problem (RRDP). Such problem can be solved by finding the optimal distribution of the RFID readers which fulfils the set of RFID planning objectives. A comparison is held between the BBO-SA algorithm and other optimization algorithms on a large RFID model. Simulation results verified the superiority of the algorithm over the compared ones for solving the RRDP with satisfying the deployment objectives.
Keywords—BBO, Benchmark functions, RFID, RRDP and SA.

1 Introduction

In the recent years, many researchers have been attracted to solve complex problems using optimization algorithms. Optimization algorithms are used to find a tradeoff among a combination of conflicting objectives. The nature can be considered as the most ordered system at all. Imitating this scheme in developing meta-heuristic optimization algorithms has become an area of great interest as it opens a wide scope of potential applications in solving complex engineering problems. These algorithms include Artificial Bee Colony (ABC) [1-4], Ant Colony Optimization (ACO) [5-12]. Particle Swarm Optimization (PSO) [13], Frog leaping (FL), Cuckoo Search (CS), Flower Pollination and Biogeography Based Optimization (BBO) [14] are also among the nature inspired algorithms. Besides these algorithms there is also the Simulated Annealing (SA) algorithm which immitates the annealing process of metals. Hybridization of algorithms are a method in which multiple algorithms run jointly and complement each other in order to generate a advantageous cooperation from their integration, known as hybrid metaheuristics [15, 16].

BBO algorithm prove itself as an efficient optimization algorithm but with some drawbacks [17]. Therefore many hybridization with BBO arises in the literature such as the hybrid technique introduced in [18] to enhance the classical BBO, by using the tabu search algorithm instead of the regular mutation operator. On the other hand, BBO-DE algorithm is introduced in [19], and aims to build a constrained variation of BBO algorithm. BBO also hybridized with Genetics Algorithms [20], ant-colony optimization (BBO-CO) [21], particle swarm optimization (BBO-PSO) [22], and recently with. invasive weed algorithm [23]. All these variations usually targeted specific problem in order to enhance its solution.

This paper suggests an optimization efficient algorithm by the hybridization of the BBO algorithm and the SA algorithm. The proposed algorithm is denoted as BBO-SA. The proposed algorithm uses the concepts of the SA to enhance the diversity of BBO solutions which in turn improves the obtained solution. The motivation to that hybridization is that the SA algorithm has the ability to escape from local optima which when integrated into the fabulous capabilities of the BBO algorithm leads to efficient searching methodology. Also, sometimes the BBO stuck in certain problems hence, the SA can be used to improve the searching capability of the BBO algorithm. Such improvement occurs due to the perturbation mechanism of the SA algorithm that helps the BBO to explore the search space in an efficient manner. The prerequisites stated by the No Free Lunch (NFL) theorem regarding the judgment on the meta-heuristic algorithms do not take into their account the complexities in these algorithms [24]. Optimization algorithms are not problem-specific such that they are not figured for particular applications. These algorithms only describe the way followed in conducting the search process. Consequently, it is not fair to compare meta-heuristic algorithms based on traditional criteria such as time complexity and memory usage. The NFL theorem defines a more suitable criterion called black-box complexity. This theory assumes the number of Function Evaluations (FES) used by the algorithm in solving the problem as the only criterion. Since there is no efficient algorithm to solve all problems, then it is not fair to judge the proposed algorithm by evaluating its performance in solving a specific problem. Therefore, the performance of the proposed algorithm in solving a number of problems must be evaluated and judged on its average performance in the solution. Consequently, the contributions in this paper are: (1) Developing an effective optimization algorithm for solving complex problems. (2) Validating the performance of the BBO-SA algorithm is by testing it in solving a set of standard 13 benchmark functions listed in [25]. Seven functions of them are unimodal functions and the others are multimodal function. Unimodal functions calculate the convergence speed of the algorithm while multimodal functions evaluate the capability of the algorithm to escape from local optima. (3) The BBO-SA algorithm is applied to solve the RRDP as complementary part to our previous work [26] which proves superiority of the BBO algorithm in solving the RRDP.

Comparisons were carried out on a working area of varying area sizes and varying loads. Optimal coverage, minimized collision, minimized number of readers and balanced load are taken into consideration in formulating the cost function that evaluates the efficiency of the proposed algorithm in satisfying the RRDP objectives. The obtained results of the BBO-SA algorithm are compared with PSO-SA algorithm and other recent bio-inspired optimization algorithms.

The paper is structured as follows: Section 2 describes the BBO-SA algorithm. The performance validation of the BBO-SA on bench mark functions is given in Section 3. The solution of the RRDP using the BBO-SA algorithm is given in Section 4. The results and discussions are shown Section 5. The conclusion is given in Section 6.

2 Proposed BBO-SA Algorithm

2.1 BBO

Before explaining the BBO-SA algorithm, we will briefly describe the searching process of both BBO and SA algorithms. The BBO is an efficient optimization algorithm that emulates biological distribution of organisms on the nature. It was first developed by Simon [17]. The BBO performs by employing a set of population individuals called habitats. Each habitat searches for a solution to the problem under consideration. A migration mechanism among these solutions is accomplished. Such migration allows the sharing of features among solutions and helps boor solutions to be enhanced. The migration among solutions is performed based on the immigration rate η and emigration rates λ of each solution.

2.2 SA

The SA starts from a random solution and performs a set of perturbations according to a set of parameters such as the initial temperature, temperature decrement step and the stopping criterion. In the initialization step the controlling parameters (i.e. the number of iterations t^{max}, initial temperature T_i, and final temperature T_f) are set and adjusted. Then each solution in the population is perturbed by applying a transition into the neighborhood of that solution to reach a new one.

There are many perturbation methods which describe the manner in which the solution parameters will be disturbed in order to construct the new solution parameters. The cost of the new solutions is evaluated and accepted if its cost value is lower than the previous solution. If the new solution has lower cost than the old one, then the new solution is accepted with a certain probability P defined as:

$$P = e^{-\frac{\Delta f}{T}} \quad (1)$$

Where Δf represents the difference in the fitness values of the two solutions and T denotes the SA temperature. The process is ended when reaching the stopping criteria (i.e. final temperature or max number of iterations is reached).

2.3 BBO-SA

The proposed BBO-SA algorithm utilizes the BBO as a basic algorithm within which the SA algorithm will be incorporated. The BBO-SA starts with the initialization of the BBO parameters such as the population sizeN_{pop}, mutation probabilityPmute, the keep parameter Nkeep which defines the number of ellit solutions. The SA parameters such as the initial temperature (T), the temperature reduction rate (α) and the number of moves (Nmoves) are also initialized. The BBO search process starts with the migration of the habitats by which the features are shared. The BBO solutions are then perturbed according to the predefined number of SA perturbations (nMove). After the perturbation, new solutions arise, and accepted or rejected based on a certain probability. Good solutions (BBO solutions that improved after the SA perturbations) are accepted while bad solutions (BBO solutions that did not improve after the SA perturbations) are accepted with a probability. Then the BBO mutation operation is applied in order to provide diversity in the output solutions. At the end of each iteration the output solutions are sorted and clipped to limit the number of solutions toN_{pop}. The search process ends when the predefined number of iterations is reached. Flowchart of the proposed algorithm is shown in Fig. 1.

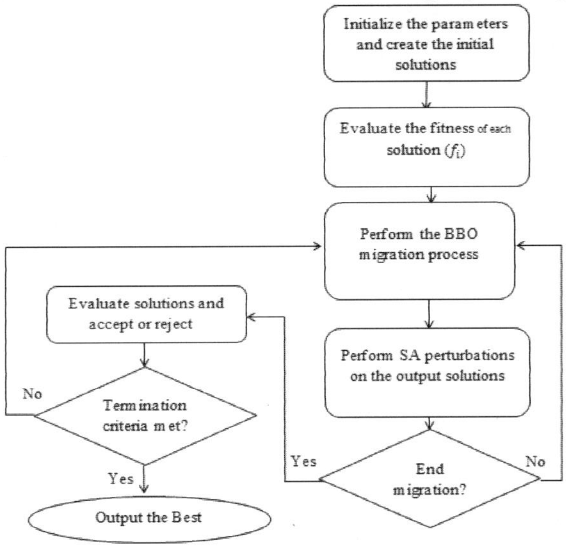

Figure 1: flowchart of the proposed BBO-SA algorithm.

3 PERFORMANCE VALIDATION OF BBO-SA

In this section we evaluated the performance of the proposed BBO-SA algorithm on solving a set of standard 13 benchmark functions listed in [25]. Seven functions of them are unimodal functions and the others are multimodal function. A comparison is held between the proposed BBO-SA and the BBO algorithm. Comparisons are conducted for 50 times with both algorithms on all functions. The performance is evaluated based on the values of the Mean, Best, Worst and standard deviation of all runs.

Table 1: The results of comparison between the BBO and BBO-SA algorithms on unimodal functions

Uni-modal functions	Algorithm		
	BBO	BBO-SA	
		$Nmove^{max} = 5$	$Nmove^{max} = 10$
f_1			
Best	1.50	3.17e-22	7.77e-11
Worst	8.56	7.38e-8	3.24e-3
Mean	4.11	3.30e-9	2.07e-5
Standard deviation	1.62	1.23e-8	7.71e-7
f_2			
Best	5.47e-1	6.77e.-1	2.48
Worst	1.28	2.36	5.92
Mean	9.26e-1	1.31	4.06
Standard deviation	1.50e-1	3.23e-1	8.98e-1
f_3			
Best	4.20e+2	2.41e-2	3.31e-13
Worst	1.80e+3	1.37e+2	3.05e-5
Mean	1.02e+3	1.09e+1	1.34e-6
Standard deviation	3.09e+2	2.09e+1	4.57e-6
f_4			
Best	3.07	7.03e-6	3.93e-4
Worst	5.46	5.23e-4	3.08e-2
Mean	4.40	9.63e-5	1.65e-2
Standard deviation	6.18e-1	1.09e-4	1.64e-2
f_5			
Best	6.68e+1	5.67e-2	9.10e-2
Worst	1.15e+3	1.39e+2	2.01e+2
Mean	2.89e+2	3.15e+1	5.83e+1
Standard deviation	2.01e+2	2.45e+1	5.88e+1
f_6			
Best	1	0	0
Worst	9	0	0
Mean	4.51	0	0
Standard deviation	1.83	0	0
f_7			
Best	1.20e-2	4.25e-3	5.73e-4
Worst	5.08e-2	3.05e-2	4.59e-3
Mean	2.43e-2	1.26e-2	2.21e-3
Standard deviation	8.23e-3	5.78e-3	8.97e-4

The process is repeated for all functions. Best, and reverse the minimum and maximum obtained costs over the 50 runs. The average value (Mean) and standard deviation (SD) are calculated from the costs over all runs. The tested unimodal functions are listed in table 5. The set of tested multimodal functions are listed in table 6. The BBO-SA algorithm is tested at different number of SA perturbations. The evaluation results on the unimodal and multimodal functions are shown in table 1 and table 2 respectively. For each function, the optimal value that the algorithm tries to reach is zero. So the algorithms are validated based on how far the algorithm from the optimal value. The results listed in table 1 and table 2 verify that the BBO-SA algorithm outperformed the BBO algorithm by about 71.4% and 83.3% in solving uni-modal and multimodal functions respectively.

Table 2: The results of comparison between the BBO and BBO-SA algorithms on multimodal functions

Multi-modal functions	Algorithm		
	BBO	BBO-SA	
		$Nmove^{max}=5$	$Nmove^{max}=10$
f_8			
Best	1.20e-2	4.25e-3	5.73e-4
Worst	5.08e-2	3.05e-2	4.59e-3
Mean	2.43e-2	1.26e-2	2.21e-3
Standard deviation	8.23e-3	5.78e-3	8.97e-4
f_9			
Best	1.09	5	1.40e+1
Worst	4.74	1.59e+1	2.82e+1
Mean	2.67	9.59	2.05e+1
Standard deviation	7.79e-1	2.11	3.02
f_{10}			
Best	6.25e-1	1.04e-1	2.23e-1
Worst	1.80	1.73	1.78
Mean	1.25	7.66e-1	8.82e-1
Standard deviation	2.29e-1	3.23e-1	2.77e-1
f_{11}			
Best	1.02	3.21e-1	1.10
Worst	1.08	1.07	1.66
Mean	1.04	8.05e-1	1.02
Standard deviation	1.48e-2	2e-1	1.17e-1
f_{12}			
Best	3.20e-3	5.01e-24	5.75e-12
Worst	5.07e-2	3.78e-9	3.13e-5
Mean	1.05e-2	1.35e-10	3.16e-6
Standard deviation	8.08e-2	6.04e-10	7.28e-6
f_{13}			
Best	7.97e-2	7.98e-25	4.22e-12
Worst	3.12e-1	2.19e-9	4.64e-3
Mean	1.88e-1	7.93e-11	2.33e-4
Standard deviation	5.01e-2	3.86e-10	7.83e-4

4 Proposed BBO-SA Algorithm for Solving The RRDP

After validating the performance of the proposed BBO-SA algorithm, we applied it for solving the RRDP problem on the RFID network model we presented in our previous work [26].

Also we hybridized the BBO algorithm with the invasive weed algorithm and introduced a self - learning methodology into an efficient algorithm called SLIWMBBO and used it for solving the RRDP in [23]. In this paper we hybridized both (the BBO with SA) and (the PSO with the SA) since the PSO is a common algorithm in solving the RRDP problem.

4.1 Problem formulation

In our model we divided the working area into a number of square grids. Readers are allowed to be positioned at the corners of the grids. Interrogation zones of the reader are considered to be a circle of a radius (r), hence the grids are of ($r/\sqrt{2}$) side lengths. The BBO-SA is applied on a working area of 36 and 64 grids with varying tag sets ranging from 200 to 400 and 600 tags. The tags is distributed randomly on the working area and marked by stars. The minimum number of RFID readers, coverage efficiency, reduced collisions and a balanced load are the deployment objectives that the algorithm tries to optimize. These objectives are formulated as

4.1.1 Minimum number of readers as:

$$\sum_{z=0}^{N_R-1} x_z \qquad (2)$$

4.1.2 Coverage Efficiency

The coverage efficiency is defined as the ratio of the number of covered grids to the total number of grids as:

$$C_E = \frac{N_{CG}}{N_G} \qquad (3)$$

4.1.3 Reduced collision

The reader collision R_C is defined as the ratio of the number of repeated covered to the total number of grids as:

$$R_C = \frac{N_{RCG}}{N_G} \qquad (4)$$

4.1.4 Balanced Load

Load balance is defined as:

$$L_B = \sum_{1}^{N_G} \frac{N_T}{N_R} \qquad (5)$$

Where N_T is the number of tags within the grid, and N_R is the number of readers covering the grid.
The combined cost function F_C is formulated as follow:

$$F_C = W_{RC} \times R_C + \frac{R \times W_R}{N_G} + W_{CE} \times Violation + \frac{L_B \times W_L}{N_G} \qquad (6)$$

The weights; W_{RC} and $W_{CE,}$ denote the weights for reader collision, coverage efficiency, respectively. While, W_R, *and* W_L are the number of readers and the load balance weights, respectively. According to RRDP priority, those weights can be varied to highlight one objective over the others. The violation measures how far the obtained coverage from the optimal coverage and is given by:

$$Violation = 1 - C_E \qquad (7)$$

$$W_{RC} + W_{CE} + W_R + W_L = 1$$

4.2 BBO-SA based solution

Once the formulation of the RRDP problem completed, the BBO-SA algorithm is employed to solve the problem and tries to find the minimum value of the fitness function (cost function) defined in equation (6). The pseudo code for BBO-SA based RRDP solution is shown in Fig. 2.

5 Simulation results and discussions

The ACO, ABC, PSO, PSO-SA, BBO and BBO-SA are all implemented for solving the RRDP. The simulations were done at model sizes of (100, and 225 grids) each for 1000 iterations. The Tags set TS (total number of tags) is changed from 200 to 400 to 600. The cost curves are shown in Figs. 3 and 4. Figure 3 provides the solution to the RRDP problem using the proposed BBO-SA algorithm as compared to some of the common algorithms for RRDP problem solving. Simulation results showed that the BBO-SA outperformed the compared algorithms in solving the RRDP on a model of 100 grids even with different tags sets

```
Define nMove which is the number of acceptedtransitions below a thresholdlevel
Initialize SA control Parameter T
Define Habitat
H(P,x) = {(P_1.X,P_1.Y,x_1),··· (P_k.X,P_k.Y,x_k),··· (P_R.X,P_R.Y,x_R)}
Create Initial Solutions
   Create and initialize R one-dimension habitats: H_i ,i ∈ [1,R]
   Evaluate initial solution cost E(H)
Repeat:
   Choose H_i with probability proportional to v_i
   if H_i is chosen then
      Calculate Emigration Probabilities
      for s = 1 to R
         Choose H_s with probability proportional to τ_s
         if H_s is chosen then
                  H_i = αH_i + (1 − α)H_s
         end if
         for m = 1 to nMove
               Apply random perturbations to the habitat H_i to the new state
               H_i = H_i + ΔH_i.
         end for
      end for
   end if
   Evaluate solution:
      ΔE(H_i) = E(H_i + ΔH_i) − E(H_i)
      if e^(ΔE(H_i)/T) ≤ rand then
            Accept the new solution
      end if
   if rand < μ
      Update the mutation
      H_i = H_i + (max(H_i) − min(H_i)) * rand
   end if
   Update the SA control Parameter T
Until stopping conditions are satisfied
```

Figure 2: Pseudo code of BBO-SA based RRDP solution

Figure 4 provides the simulation results for a large model of 225 grids at variant tag sets. It can be noticed from the figure that the BBO-SA algorithm outperformed the compared algorithms in solving the RRDP problem. The simulation results are concluded in table 3. It can be seen that the BBO-SA achieves the lowest optimization cost and the minimum number of deployed readers than the compared algorithms.

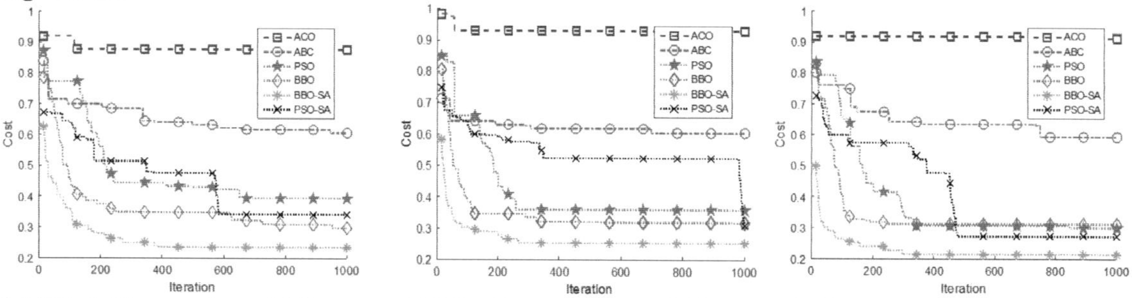

(a) TS =200 tags, $N_G = 100$ (b) TS =400 tags, $N_G = 100$ (c) TS =600 tags, $N_G = 100$

Figure 3: Cost curves for the 100 grids model at different loads: (a) 200, (b) 400 and and (c) 600.

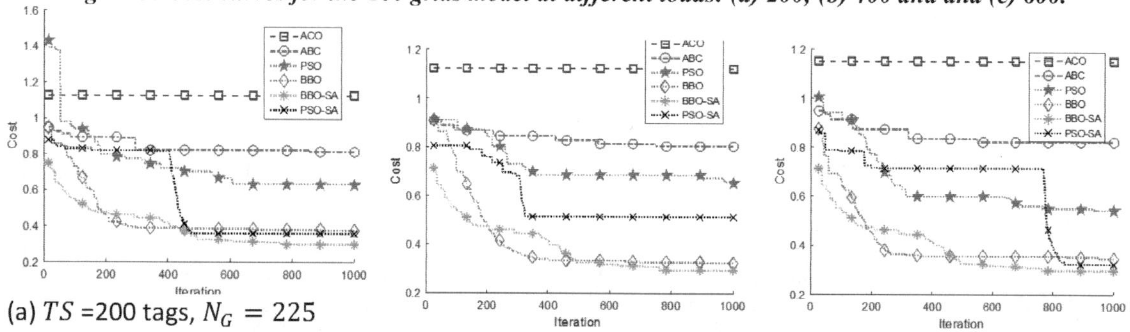

(a) TS =200 tags, $N_G = 225$ (b) TS =400 tags, $N_G = 225$ (c) TS =600 tags, $N_G = 225$

Figure 4: Cost curves for the 225 grids model at different loads:(a) 200, (b) 400 and (c) 600 tags.

Table 3: The results of the compared optimization algorithms on an area of, 100 and 225 grids for a tag set of 200 tags.

Optimization Algorithm	$N_G = 100$		$N_G = 225$	
	R	(O_C)	R	(O_C)
ACO	64	0.877	159	1.125
ABC	47	0.607	125	0.814
PSO	38	0.394	109	0.628
BBO	34	0.299	73	0.310
PSO-SA	39	0.3412	82	0.359
Proposed BBO-SA	25	0.2125	71	0.297

6 Conclusion

This paper proposes an efficient meta-heuristic optimization algorithm developed by hybridizing the BBO algorithm with the SA algorithm into an algorithm called BBO-SA. For validating the performance of the proposed algorithm, we tested its performance in solving benchmark functions.

Validation results showed that the BBO-SA algorithm is superior to the BBO algorithm in solving both unimodal and multimodal functions. After validating its performance, the proposed BBO-SA algorithm is compared to other bio-inspired algorithms (i.e. ABC, ACO, PSO, BBO, and PSO-SA) for solving the RRDP problem on a working area of varying area sizes and different loads. The results of comparison proved that the proposed BBO-SA based RRDP solution outperformed the compared algorithms in locating the optimal positions of the RFID readers with a tradeoff among all the RRDP objectives. The superiority of the proposed BBO-SA algorithm returns to the proposed hybridization methodology that integrates the advantages of both BBO and SA algorithm. The improved performance of the BBO-SA algorithm makes it suitable for solving the RRDP problem on large scales.

References

[1] Kang F, Li J and Li H 2013 Artificial bee colony algorithm and pattern search hybridized for global optimization Applied Soft Computing 13 1781-91
[2] Karaboga D and Akay B 2009 A comparative study of artificial bee colony algorithm Applied mathematics and computation 214 108-32
[3] Karaboga D and Basturk B 2007 Artificial bee colony (ABC) optimization algorithm for solving constrained optimization problems. In: International Fuzzy Systems Association World Congress: Springer) pp 789-98
[4] Zhu G and Kwong S 2010 Gbest-guided artificial bee colony algorithm for numerical function optimization Applied Mathematics and Computation 217 3166-73
[5] Hsiao Y-T, Chuang C-L and Chien C-C 2004 Ant colony optimization for best path planning. In: Communications and Information Technology, 2004. ISCIT 2004. IEEE International Symposium on: IEEE) pp 109-13
[6] Dorigo M, Birattari M, Blum C, Clerc M, Stützle T and Winfield A 2008 Ant Colony Optimization and Swarm Intelligence: 6th International Conference, ANTS 2008, Brussels, Belgium, September 22-24, 2008, Proceedings vol 5217: Springer)
[7] Dorigo M and Gambardella L M 1997 Ant colonies for the travelling salesman problem biosystems 43 73-81
[8] Fan X, Luo X, Yi S, Yang S and Zhang H 2003 Optimal path planning for mobile robots based on intensified ant colony optimization algorithm. In: Robotics, Intelligent Systems and Signal Processing, 2003. Proceedings. 2003 IEEE International Conference on: IEEE) pp 131-6
[9] Fidanova S 2007 Ant colony optimization and multiple knapsack problem Handbook of research on nature inspired computing for economics and management 498-509
[10] Lewis A, Weis G, Randall M, Galehdar A and Thiel D 2009 Optimising efficiency and gain of small meander line RFID antennas using ant colony system. In: 2009 IEEE Congress on Evolutionary Computation: IEEE) pp 1486-92
[11] Parsons S 2005 Ant Colony Optimization by Marco Dorigo and Thomas Stützle, MIT Press, 305 pp., $40.00, ISBN 0-262-04219-3 The Knowledge Engineering Review 20 92
[12] Slyusar V I and Ermolaev S 2009 Antenna synthesis based on the ant colony optimization algorithm
[13] Garg H 2016 A hybrid PSO-GA algorithm for constrained optimization problems Applied Mathematics and Computation 274 292-305
[14] Kar A K 2016 Bio inspired computing–A review of algorithms and scope of applications Expert Systems with Applications 59 20-32
[15] Jena U, Das P and Kabat M 2020 Hybridization of meta-heuristic algorithm for load balancing in cloud computing environment Journal of King Saud University-Computer and Information Sciences
[16] Ting T, Yang X-S, Cheng S and Huang K 2015 Hybrid metaheuristic algorithms: past, present, and future Recent advances in swarm intelligence and evolutionary computation 71-83
[17] Simon D 2008 Biogeography-based optimization IEEE transactions on evolutionary computation 12 702-13
[18] Lim W L, Wibowo A, Desa M I and Haron H 2016 A biogeography-based optimization algorithm hybridized with tabu search for the quadratic assignment problem Computational intelligence and neuroscience 2016
[19] Boussaïd I, Chatterjee A, Siarry P and Ahmed-Nacer M 2011 Hybridizing biogeography-based optimization with differential evolution for optimal power allocation in wireless sensor networks IEEE Transactions on Vehicular Technology 60 2347-53
[20] Li W, Guo W, Wang L, Kang Q and Wu Q 2013 Hybrid BBO and GA algorithms based on elites operation International Journal of Modelling, Identification and Control 18 9-17

[21] Dai Z, Ma Q, Zhao D and Yan S 2020 A novel hybrid optimization algorithm combined with BBO and ACO. In: 2020 39th Chinese Control Conference (CCC): IEEE) pp 1581-6
[22] Cheng G, Lv C, Yan S and Xu L 2016 A novel hybrid optimization algorithm combined with BBO and PSO. In: 2016 Chinese Control and Decision Conference (CCDC): IEEE) pp 1198-202
[23] Zahran E G, Arafa A A, Saleh H I and Dessouky M I 2020 A self learned invasive weed-mixed biogeography based optimization algorithm for RFID network planning Wireless Networks 26 4109-27
[24] Wolpert D H and Macready W G 1997 No free lunch theorems for optimization IEEE transactions on evolutionary computation 1 67-82
[25] Jaballah A and Meddeb A 2017 A new variant of cuckoo search algorithm with self adaptive parameters to solve complex RFID network planning problem Wireless Networks 1-20
[26] Zahran E, Arafa A, Saleh H and Dessouky M 2018 Biogeography Based Optimization Algorithm for Efficient RFID Reader Deployment. In: 2018 13th International Conference on Computer Engineering and Systems (ICCES): IEEE) pp 454-9

Application of Suggested Three Parameters Collective Rotational Model to Superdeformed Thallium Odd-Mass Nuclei

A.M. Khalaf [1], M.M. Taha [2], A.M. Ismail [3]

[1] Physics Department, Faculty of Science, Al-Azhar University, Cairo, Egypt.
[2] Mathematics and Theoretical Physics Department, Nuclear Research Center, Egyptian Atomic Energy Authority, Cairo, Egypt.
[3] Department of Radioactive Waste Management, Hot Laboratories and Waste Management Center, Egyptian Atomic Energy Authority, Cairo,Egypt.

Email: mahmoudmt@hotmail.com

Abstract. The superdeformed rotational bands (SDRB's) of axial symmetric Tl odd-A nuclei are described within the framework of the suggested model as an extended versionof Bohr-Moltelson (BM) model. The suggested model has three parameters rather than the BM model's two parameters. The model is employed to calculated the transition energies and therefore the unknown level spins within the three pairs of signature partners ^{191}Tl (SD1, SD2), ^{193}Tl (SD1, SD2) and ^{195}Tl (SD1, SD2). The optimized model parameters and therefore the unknown spins for every band are adjusted by employing a computer simulated search program to induce a minimumroot mean square deviation between the calculated theoretical and experimental transition energies $E_\gamma(I)$. The most effective adopted model parameters and spins are wont to calculate the kinematic $J^{(1)}$ and dynamic $J^{(2)}$ moments of inertia, and rotational frequency $\hbar\omega$,. The calculated results agree excellently with the experimental ones. The behaviour of $J^{(1)}$ and $J^{(2)}$ with $\hbar\omega$ is examined and discussed. Also, the calculated transition energies and spins are accustomed investigate and exhibit the $\Delta I = 1$ staggering effects in transition energies between each pair of the signature partners. This can be done by considering three staggering indices looking on the dipole γ-transitions between the 2 bands of the signature partners and therefore the quadrupole γ-transitions within each band. All the studied three signature partner pairs showed an oversized amplitude staggering pattern. The $\Delta I = 1$ energy staggering also appears within the transition energies after subtracting a rigid rotor reference when plotted versus spin. The phenomenon of identical bands (IB's) is additionally investigated within the two pairs [^{193}Tl (SD1), ^{195}Tl (SD1)] and [^{193}Tl (SD2), ^{195}Tl (SD2)]by using the concept of incremental alignment.

1. Introduction

Superdeformed (SD) nuclei have become one of the most exciting and attention-grabbing areas of theoretical nuclear research since the observation of the first superdeformed rotational bands (SDRBs) in ^{152}Dy [1] and in ^{191}Hg [2]. Now a great number of axially symmetric SDRBs have been identified

Journal of Physics: Conference Series **2304** (2022) 012014 doi:10.1088/1742-6596/2304/1/012014

in different mass regions: A~ 60, 80, 130, 150, and 190 [3]. In addition, nuclei are ascribed to having triaxial SD shapes [4- 6].

The difference between the SDRBs in many mass regions is manifested in the behaviour of the dynamical moment of inertia $J^{(2)}$. The A~ 190 mass region is of special interest since in this region the SDRB's were taken notice that down to quite low spin and display the same smooth rise in $J^{(2)}$ with increasing rotational frequency $\hbar\omega$ [7], this is due to the fact that nucleons gradually alignment and occupy high intruder orbitals with diminishing pair correlation [8, 9].

For all SDRBs, the transition energies are the only quantities to be detected; spins, parities, and excitation energies are unknown due to the difficulty of establishing the excitation of the SD band on the ground states. Several theoretical procedures for assigning spins to SD states were proposed [10-14]. The $\Delta I = 2$ staggering, also known as the $\Delta I = 4$ bifurcation, is an exciting development for nuclear structure studies. It corresponds to an alternating shift in transition energies between two sequences of states with spins I, I+4, I+8,... and I+2, I+6, I+10,... within the SDRBs [15, 16, 17, 18].

It was found that many SDRBs in odd-A nuclei in the A ~ 190 mass region is signature partner SD bands [19-23]. Most of these signature partner pairs exhibit a $\Delta I = 1$ staggering effect in their transition energies [24-27]. For two signature partner bands in the same nucleus, the levels with angular momenta I, I+2, I+4,... in the first band are displaced relative to the levels with angular momenta I+1, I+3, I+5,... in the other band. A spin-dependent zigzag behaviour was observed, which was unexpected in the transition energies of SD nuclei. The intrinsic structure and the bandhead moments of inertia of these signature partners are found to be almost identical. The $\Delta I = 1$ staggering effect is also investigated in normal deformed nuclei between the energies of states of the ground and octapole bands [28].

One of the unexpected features that happen in SDRBs is the presence of identical bands (IB's) [25, 27, 29- 31]. This interested discovery of IB's has been seen also in pairs of normal deformed nuclei [32, 33].

2. Outline of the suggested collective rotational model

The rigid rotor formula gives the rotational ground state energy E(I) versus spin I for an axially symmetric nucleus in the strong coupling limit,

$$E(I) = AI(I + 1) \tag{1}$$

where the rotational parameter A is the reciprocal of the nuclear moment of inertia $A = \hbar/2J$. For high in this spin state, deviation between the rigid rotor formula (1) and experimental data occurs.

Bohr and Moltelson (BM) [34] expanded E(I) as an infinite power series of I(I+1). The two-term BM formula is as follows:

$$\begin{aligned} E(I) &= \alpha (I + 1) I + \beta ((I + 1) I)^2 \\ &= \alpha (I + 1) I [1 + \gamma (I + 1) I] \end{aligned} \tag{2}$$

using $\gamma = \beta/\alpha$. According to variable moment of inertia (VMI) model [35], there is a gradual increase in the moment of inertia J with increasing spin I.

We suggest that

$$\alpha = \frac{\hbar^2}{2J} \rightarrow \frac{\hbar}{2J\,[1 + \sigma\, I(I + 1)]} \tag{3}$$

Therefore, the extended BM two-parameter formula with three parameters, J, γ and σ reads:

$$E(I) = \frac{\hbar^2}{2J}\,(I + 1)I\,\frac{1 + \gamma\, I(I + 1)}{1+\sigma\, I(I + 1)} \tag{4}$$

This formula is the most general formula, because:

- **i** When $\sigma = \gamma$ it gives the pure rigid rotor equation (1)
- **ii** When $\sigma = 0$ it gives the two parameters BM formula equation (2)
- **iii** When $\gamma = 0$ it gives the soft rotor model [30- 32]

$$E(I) = \frac{\hbar^2}{2J} (I + 1) \frac{I}{1+\sigma I(I+1)} \tag{5}$$

Within the framework of the collective rotation model, $J^{(1)}$, $J^{(2)}$ and $\hbar\omega$ for our suggested model equation (4), taking $I(I+1) = \hat{I}^2$ reads

$$J^{(1)}(I) = \hbar^2 \hat{I} \left(\frac{dE(\hat{I})}{d\hat{I}}\right)^{-1} = \frac{\hbar^2(1+\sigma \hat{I}^2)^2}{2\alpha[1+\gamma \hat{I}^2(2+\sigma \hat{I}^2)]} \tag{6}$$

$$J^{(2)}(I) = \hbar^2 \left(\frac{d^2E(\hat{I})}{d\hat{I}^2}\right)^{-1} = \frac{\hbar^2(1+\sigma \hat{I}^2)^3}{2\alpha[(1+6\gamma \hat{I}^2)+\sigma \hat{I}^2(3\gamma \hat{I}^2-\sigma\gamma \hat{I}^3)]} \tag{7}$$

$$\hbar\omega(I) = \frac{dE(\hat{I})}{d\hat{I}} \tag{8}$$

$$= 2\alpha \hat{I} \frac{1+\gamma \hat{I}^2(2+\sigma \hat{I}^2)}{(1+\sigma \hat{I}^2)^2} \tag{9}$$

For special case when $\sigma = 0$, yield

$$J^{(1)}(I) = \frac{\hbar^2}{2\alpha(1 + 2\gamma \hat{I}^2)} \tag{10}$$

$$J^{(2)}(I) = \frac{\hbar^2}{2\alpha(1 + 6\gamma \hat{I}^2)} \tag{11}$$

$$\hbar\omega(I) = 2\alpha \hat{I} (1 + 2\gamma \hat{I}^2) \tag{12}$$

which are the same quantities for two parameters BM formula with $\gamma = \beta/\alpha$.

Experimentally, for SDRBs, only γ-ray transition energies are available and are commonly translated into values of $\hbar\omega$ and $J^{(2)}$ as:

$$\hbar\omega(I) = \frac{1}{4}[E_\gamma(I + 2) + E_\gamma(I)] \quad (\text{MeV}) \tag{13}$$

$$J^{(2)}(I) = \frac{4}{E_\gamma(I+2)-E_\gamma(I)} \quad (\hbar^2\text{MeV}^{-1}) \tag{14}$$

Where,

$$E_\gamma(I) = E(I) - E(I - 2) \tag{15}$$

Also, after knowing the spins, $J^{(1)}$ is givenby

$$J^{(1)}(I) = \frac{2I+1}{E_\gamma(I)} \tag{16}$$

It is seen that, while $J^{(1)}$ depends on spin I, the $J^{(2)}$ does not.

3. The ΔI = 1 staggering in signature partner pairs of odd-mass SD nuclei.

To investigate and exhibit the ΔI = 1 staggering in the transition energies of signature partner pairs of odd-mass nuclei using our suggested three-parameters model, we consider three different staggering indices:

i The first staggering index is a limit difference approach to the second order derivative

$\Delta^2 E_\gamma(I)$, which is determined by the variance between the rate transition energies $E_\gamma(I+2 \rightarrow I)$, $E_\gamma(I \rightarrow I-2)$ in one band and the transition energy $E_\gamma(I+1 \rightarrow I-1)$ in its signature partner, it reads:

$$\Delta^2 E_\gamma(I) = \frac{1}{2}\left[E_\gamma(I+2) + E_\gamma(I) - E_\gamma(I+1)\right]$$
$$= \frac{1}{2}\left[E_\gamma(I+2) - 2E_\gamma(I+1) + E_\gamma(I)\right] \quad (17)$$

This form contains three sequent transition energies and is indicated by the three-point formula.

ii The second staggering index depends on the $\Delta I = 1$ dipole transition energies linking the signature partners and the $\Delta I = 2$ quadrupole transition energies within each band. It reads:

$$Y(I) = \left(\frac{(4I-2)}{2I}\right)\left(\frac{E_{\gamma 1}(I)}{E_{\gamma 2}(I)}\right) - 1 \quad (18)$$

With

$$E_{\gamma 1}(I) = E(I) - E(I-1) \quad (19)$$
$$E_{\gamma 2}(I) = E(I) - E(I-2) \quad (20)$$

The Y(I) formula vanishes for pure rigid rotor.

iii The third staggering index represents the transition energies after the minus of a rotational term, that is, the transition energies dependent on a rigid rotor reference, and it reads

$$S(I) = E_{\gamma 2}(I) - A(4I-2) \quad (21)$$

4. Identical bands in SDRB's

One of the most interesting discoveries in SDRB's is the presence of IB's [25, 27, 29- 31], i.e., nearly identical transition energies Eγ of the sent out γ-radiation in several pairs of bands within 1-3 KeV over large spin intervals. The phenomenon of IB's has created much theoretical interest [36 -40]. Since the transition energy is almost equal twice the rotational frequency, it means that the rotational frequencies of the two IB's are more identical, and also denote that the dynamical moments of inertia are very nearly. There are a number of standard ways to determine whether a pair of bands is identical or not. There are a number of standard ways to define whether a pair of bands is similar or not. The most common of these ways is the comparison of E2 transition energies Eγ or the percentage differences ΔEγ/Eγ.

In the case of SD nuclei in the $A\sim$ 190 mass region, Stephens *et al* [41, 42] proposed the incremental alignment Δi, which is spin independent and depends only on the formula between the transition energy in the SD band of concern and that of a reference band. It reads:

$$\Delta i_{AB} = 2\left(\frac{\Delta E_\gamma}{\Delta E_\gamma^{\text{ref}}}\right) \quad (22)$$

with,

$$\Delta E_\gamma = E_\gamma^A(I+2) - E_\gamma^B(I) \quad (23)$$

$$\Delta E_{\gamma}^{\mathrm{ref}} = E_{\gamma}^{B_{\mathrm{ref}}}(I+2) - E_{\gamma}^{B_{\mathrm{ref}}}(I) \tag{24}$$

where A define the nucleus of concern and B is the reference nucleus. In region A~ 190 the yrast SD band in ^{192}Hg is selected as the reference because of its approaches to doubly magic character.

Table 1: The parameters of the model *J, γ* and *σ*, that fit the best were chosen for the signature partners in odd-mass SD Tl nuclei. The spin proposition values of the bandhead are given as well. The last column gives the relative rms deviation of the determined transition energies with the experimental results.

SD band	E_γ (KeV) $(I_0+2 \rightarrow I_0)$	I_0 $(\hbar)$	J $\hbar^2$ MeV^{-1}	$\gamma \times 10^5$	$\sigma \times 10^5$	rms Deviation
^{191}Tl (SD1)	276.5	11.5	92.7557	4.1545	7.91055	0.0351
^{191}Tl (SD2)	296.3	12.5	92.8177	4.7159	9.9001	0.0397
^{193}Tl (SD1)	206.6	8.5	95.6809	5.9977	10.099	0.0386
^{193}Tl (SD2)	227.3	9.5	95.6956	1.9693	5.7688	0.0311
^{195}Tl (SD1)	146.2	5.5	95.1275	6.7949	10.320	0.0410
^{195}Tl (SD2)	167.5	6.5	94.8298	8.3583	12.818	0.0606

5. Numerical calculations and discussion

In Tl nuclei, three signature partners are considered, namely ^{191}Tl (SD1, SD2), ^{193}Tl (SD1, SD2), and ^{195}Tl (SD1, SD2).

The transition energies $E\gamma(I) = E(I) - E(I-2)$ for each band can be calculated using equation (4). These transition energies are applied to fit the observed transition energies of the studied SD bands with J, γ, and σ and the bandhead spin value I_0 as free parameters. The spin of the bandhead is chosen as the nearest half integer of the I_0, and then second fit with only J, γ, and σ as free parameters is made to calculate the model parameters J, σ. To get a minimum root mean square (rms) deviation, the parameters were changed using a computer simulated search algorithm.

$$\chi = \sqrt{\frac{1}{N}\sum_{i=1}^{N}\left(\frac{E_{\gamma}^{\mathrm{exp}}(i)-E_{\gamma}^{\mathrm{cal}}(I_i)}{\delta E_{\gamma}^{\mathrm{exp}}(i)}\right)^2} \tag{25}$$

of the determined energies $E_{\gamma}^{\mathrm{cal}}$ from the experimental energies $\mathbf{E}_{\gamma}^{\mathrm{exp}}$, where N is the number of data points used in the fitting steps and denotes the experimental error in the experimental $E\gamma$ energies. The steps of fitting is repeated for several trial values of I_0, and the value of I_0 that leads to the minimum χ is chosen for spin assignment of the observed level. For each band in the ^{191,193,195}Tl nuclei, the resulting adopted best model partners J, γ, σ and the lowest bandhead spin I_0 are listed in Table (1). The experimental lowest transition energy $E\gamma(I_0 + 2 \rightarrow I_0)$ and the rms deviation are also given. The spin propositions are in agreement with previous work [34]. A comparison between the theoretical Gamma transition Energies Eγ(I) Over Spin EGOS = Eγ(I)/2I extracted from our suggested model for the three pairs of signature partner and the experimental ones are listed in Table (2). Excellent agreement has been obtained.

Table 2: Comparison between the calculated and experimental EGOS(I) = Eγ(I)/2I of SD bands of signature partner ^{191}Tl (SD1,SD2), ^{193}Tl (SD1,SD2) and ^{195}Tl (SD1,SD2)

^{191}Tl						^{193}Tl					
SD1			SD2			SD1			SD2		
EGOS (I) (KeV)			EGOS (I) (KeV)			EGOS (I) (KeV)			EGOS (I) (KeV)		
Spin	Cal.	Exp.	Spin	Cal.	Exp.	Spin	Cal.	Exp.	Spin	Cal.	Exp.
11.5	10.2504	10.2407	12.5	10.2328	10.2172	8.5	9.8707	9.8380	9.5	9.9051	9.8826
12.5	10.2582	10.2483	13.5	10.2260	10.2181	9.5	9.9940	9.9820	10.5	9.9370	9.9222
13.5	10.2482	10.2571	14.5	10.2028	10.2108	10.5	9.9292	9.9206	11.5	9.9465	9.9419
14.5	10.2241	10.2256	15.5	10.1663	10.1682	11.5	9.9246	9.9212	12.5	9.9398	9.9428
15.5	10.1883	10.1930	16.5	10.1185	10.1266	12.5	9.9052	9.9027	13.5	9.9208	9.9230
16.5	10.1425	10.1446	17.5	10.0608	10.0571	13.5	9.8742	9.8850	14.5	9.8923	9.8953
17.5	10.0878	10.0882	18.5	9.9943	9.9924	14.5	9.8338	9.8422	15.5	9.8564	9.8659
18.5	10.0252	10.0254	19.5	9.9198	9.9315	15.5	9.7858	9.7897	16.5	9.8145	9.8254
19.5	9.9554	9.9576	20.5	9.8390	9.8377	16.5	9.7315	9.7377	17.5	9.7682	9.7727
20.5	9.8790	9.8698	21.5	9.7499	9.7446	17.5	9.6723	9.6771	18.5	9.7186	9.7186
21.5	9.7965	9.7955	22.5	9.6556	9.6507	18.5	9.6093	9.6147	19.5	9.6669	9.6634
22.5	9.7085	9.7154	23.5	9.5558	9.5547	19.5	9.5435	9.5430	20.5	9.6143	9.6089
23.5	9.6153	9.6130	24.5	9.4510	9.4545	20.5	9.4761	9.4724	21.5	9.5619	9.5591
						21.5	9.4080	9.3986	22.5	9.5109	9.5093
						22.5	9.3403	9.3337	23.5	9.4623	9.4620
						23.5	9.2741	9.2753	24.5	9.4175	9.4204
						24.5	9.2106	9.2164			

^{195}Tl					
SD1			SD2		
EGOS (I) (KeV)			EGOS (I) (KeV)		
Spin	Cal.	Exp.	Spin	Cal.	Exp.
5.5	9.7771	9.7466	6.5	9.8680	9.8529
6.5	9.9019	9.9052	7.5	9.9539	9.9238
7.5	9.9698	9.9478	8.5	9.9949	9.9640
8.5	10.0037	9.9888	9.5	10.0071	9.9827
9.5	10.0148	10.0161	10.5	9.9987	9.9878
10.5	10.0092	9.9885	11.5	9.9747	9.9756
11.5	9.9908	9.9948	12.5	9.9381	9.9390
12.5	9.9620	9.9744	13.5	9.8913	9.9044
13.5	9.9252	9.9404	14.5	9.8358	9.8469
14.5	9.8798	9.8921	15.5	9.7730	9.7924
15.5	9.8288	9.8327	16.5	9.7041	9.7157
16.5	9.7724	9.7711	17.5	9.6301	9.6360
17.5	9.7113	9.7142	18.5	9.5520	9.5538
18.5	9.6464	9.6373	19.5	9.4708	9.4637
19.5	9.5783	9.5690	20.5	9.3872	9.3726
20.5	9.5077	9.5040	21.5	9.3023	9.2922
21.5	9.4351	9.4405	22.5	9.2171	9.2345
22.5	9.3975	9.3638	23.5	9.1324	9.1470

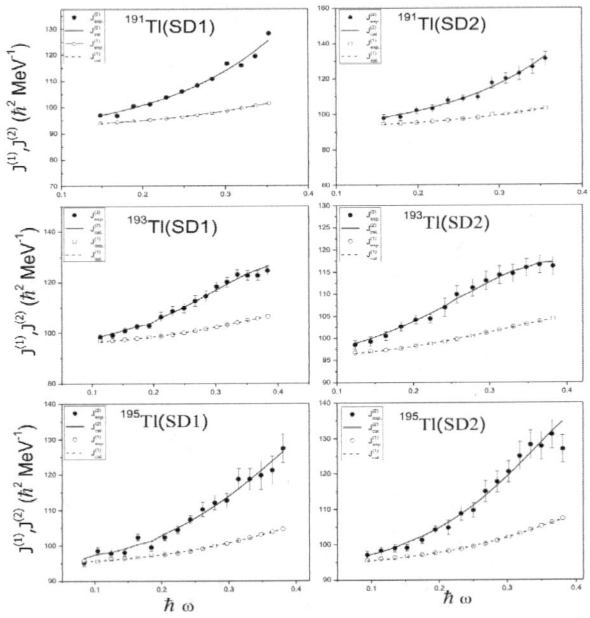

Figure 1: Values of theoretical $J^{(1)}$ and $J^{(2)}$ from the suggested model calculation as the dashed and solid smooth curves, respectively, as a function of $\hbar\omega$ for the SDRB's of signature partners in Tl nuclei. For comparison, the experimental J(1) and J(2) are shown as unlinked open and filled circles.

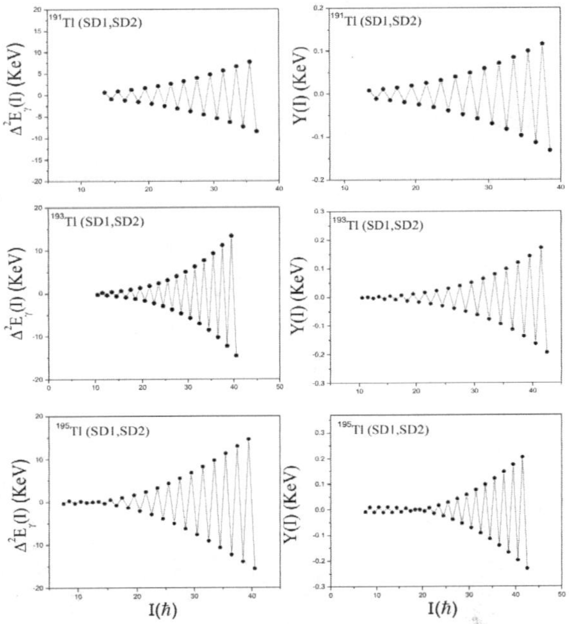

Figure 2: The calculated ΔI = 1 staggering indices $\Delta^2E\gamma(I)$ and Y(I) versus of nuclear spin I for the signature partners ^{191}Tl (SD1,SD2), ^{193}Tl (SD1,SD2) and ^{195}Tl (SD1,SD2).

The best parameters of the model are retained and used to calculate the $J^{(1)}$ and $J^{(2)}$. Figure (1) shows the behaviour of $J^{(1)}$ and $J^{(2)}$ versus $\hbar\omega$ compared with the experimental results. The observed gradual rise in $J^{(1)}$ and $J^{(2)}$ moments of inertia at intermediate rotational frequencies of 0.2 MeV $\leq$ $\hbar\omega$ $\leq$ 0.3 MeV is due to the progressive rotational alignment of a pair of high-j nucleons with pairing decreasing. A saturation of the $J^{(2)}$ moments of inertia at rotational frequency $\hbar\omega > 0.32$ MeV is seen in the two bands SD1, SD2 in ^{193}Tl which reflect the collective effects of the proton pairing blocking and the complete $j_{15/2}$ neutron alignment. From Figure (1), we see that the two pairs (^{193}Tl (SD1), ^{195}Tl (SD2)), (^{193}Tl (SD2), ^{195}Tl (SD1)) display striking similarities in $J^{(1)}$ and $J^{(2)}$ in the rotational frequency range of 0.10 MeV $< \hbar\omega <$ 0.35 MeV, these two pairs can be classified as IB's.

Now, we would like to focus on $\Delta I = 1$ energy staggering phenomenonpresent in odd-mass Tl nuclei. For our selected three pairs of signature partners in ^{191,193,195}Tl, the values of the energy shift $\Delta^2E_\gamma(I)$ equation (17) and the staggering index Y(I) equation (18) as a function of the spin I havebeen extracted and plotted in Figure (2). It is seen that the three pairs display large capacity staggering. The $\Delta^2E_\gamma(I)$ for ^{193}Tl and ^{195}Tl are remarkably similar.

Also to exhibit the staggering effects the transition energies $E_\gamma(I) = E(I) \quad E(I - 2)$ relative to a rigid rotor reference energy $\xi = 4.8(4I - 2)$KeV versus of spin for our three signature partners is displayedin Figure (3). This reference term highlights the rotational motion in the spin region and magnifies the staggering effect. It is seen that this staggering index shows a large signature splitting at higher spins. A zigzag pattern occurs from linking the closed circles (SD1) and the open circles (SD2).

About the identical bands (IB's) a plot of percentage differences in transition energies $(\Delta E_\gamma/E_\gamma) = (E_\gamma(SDa)-E_\gamma(SDb))/E_\gamma(SDa)$ where the index (a) refers to the heavier mass nuclide, versus rotational frequency $\hbar\omega$ for the two pairs of bands (^{193}Tl (SD1), ^{195}Tl (SD2)) and (^{193}Tl (SD2),

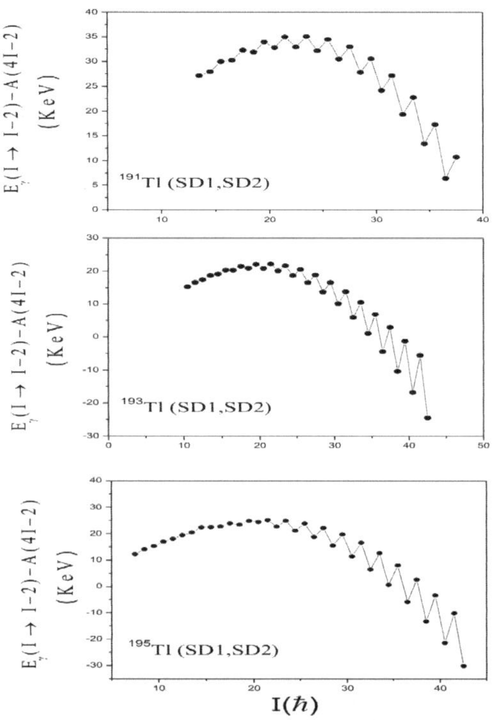

Figure 3: The ΔI = 1 staggering in the determined transition energies subtract rigid rotor reference with rotational partners A=4.8 keV versus of spin I for the signature partners 191Tl (SD1,SD2), 193Tl (SD1,SD2) and 195Tl (SD1,SD2).

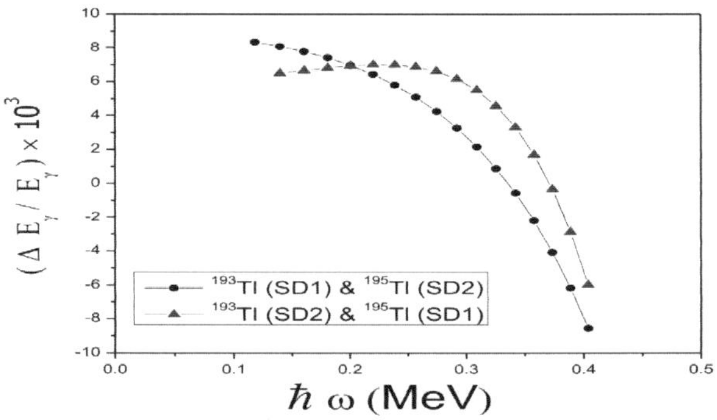

Figure 4: Percentage difference $\Delta E\gamma/E\gamma$ in transition energies versus of rotational frequency for the two pairs of identical bands (^{193}Tl (SD1), ^{195}Tl (SD2)) and (^{193}Tl (SD2), ^{195}Tl (SD1)).

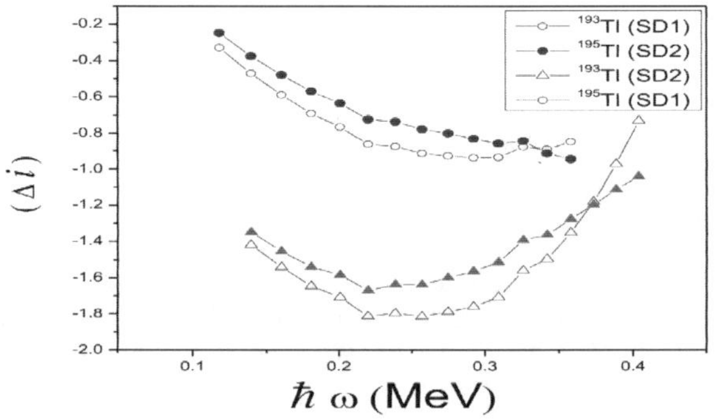

Figure 5: Incremental alignment Δi versus of rotational frequency $\hbar\omega$ for the two identical bands ((^{193}Tl (SD1), ^{195}Tl (SD2)) and (^{193}Tl (SD2), ^{195}Tl (SD1)).

^{195}Tl (SD1))is illustrated in Figure (4). We see that the percentage differences are few. Therefore, the two bands are considered IB's. It is motivating to observe that the bandhead moments of inertia J_0 of each pair are almost identical. Additional information about the IB's can be obtained by comparing the transition energies directly to those in the other bands. The incremental alignment Δi is plotted in Figure(5) relative to the reference ^{192}Hg (SD1) against rotational frequency $\hbar\omega$ for our two pairs of signature partners. The data shows a cluster around the values $\Delta I = -1$ and 0. We chased ^{192}Hg as a reference because ^{192}Hg is considered as the core for ^{193}Tl (^{192}Hg $\oplus\pi$) where π is the additional proton which occupy the third intruder orbital $\pi_{5/2}$ + [642].

6. Conclusion

In the framework of collective rotational motions, a new three-parameter model based on an extended version of the Bohr-Mottelson two-parameter model is presented and used to describe the properties of nuclear superdeformed rotational bands (SDRBs). The present model is applied to three pairs of signature partners in odd-mass Thallium nuclei. The determined level spins, transition energies, rotational frequencies, kinematic and dynamic moments of inertia are compared with experimental data. Model parameters are extracted by fitting the transition energies of the proposed model with experimental ones. The systematic variations of kinematic and dynamic moments of inertia are studied versus of the rotational frequency. The effectof the $\Delta I = 1$ energy staggering observed in SDRBs between states of signature partners differing by one unit of angular momentum in ^{191,193,195}Tl nuclei is examined by using two suggested staggering indices. The first staggering index represents the divergence between the average transition energies of $E_\gamma(I+2\rightarrow I)$, $E_\gamma(I\rightarrow I-2)$ in one band and the transition $E_\gamma(I+1\rightarrow I-1)$ in its signature partner. The second staggering index depends on the $\Delta I = 1$ transition linking the two signature partners and the $\Delta I = 2$ transition energies within each band. Moreover, the $\Delta I = 1$ staggering is exhibited by plotting the transition energies after subtracting a rigid rotor reference versus spin. All the chosen three pairs of signature partners display large capacity staggering patterns. Further applications of our model could be the exhibition of identical bands in ^{193,195}Tl nuclei by using the concept of incremental alignment.

References

[1] P.J. Twin et al, 1986, "Observation of a Discrete Line Super-deformed Band up to 60$\hbar$ in ^{152}Dy", Phys. Rev. Let., Vol. 57, No. 7, pp. 811-814.

[2] E.F. Moore et al, 1989, "Observation of super-deformation in ^{191}Hg", Phys. Rev. Let., Vol. 63, No. 4, pp. 360-363.

[3] http://www.nndc.bnl.gov/ (ENSDF database).

[4] H. Schrack Petersen et al, 2007, "Super-deformed triaxial bands in ^{163,165}Lu", Nucl. Phys., Vol. 594, No. 2, pp. 175-202.

[5] Y.C. Zhang et al, 2007, "Nuclear shapes of highly deformed bands in ^{171,172}Hf and neighboring Hf isotopes",Phys. Rev. C, Vol. 76, No. 6, pp. 064321.

[6] H.A. Ghanim et al, 2021, "Features of Triaxial Superdeformed Bands in Odd Mass Nuclei Using the Cranked Nilsson Strutinsky Model", Phys. Atomic Nucl., Vol. 84, No. 4, pp. 433-440.

[7] R.V.E. Jensseens and T.L. Khoo, 1991, "Superdeformed Nuclei", Anna. Rev. Nucl. Part. Sci., Vol. 41, pp. 321-355.

[8] D. Ye, et al, 1990, "Superdeformed band in ^{192}Hg", Phys. Rev. C, Vol. 41, No. 1, pp. R13.

[9] M.A. Riley et al, 1990, "Multiple superdeformed bands in 194Hg and their dynamical moments of inertia",Nucl. Phys. A, Vol. 512, No. 1, pp. 178-188.

[10] J.A. Beacker et al, 1990, "Level spin and moments of inertia in superdeformed nuclei near A = 194", Nucl. Phys. A, Vol. 520, pp. c187-c194

[11] J.Y. Zeng et al, 1991, "Spin determination and quantized alignment in the superdeformed bands in ^{152}Dy, ^{151}Tb, and ^{150}Gd", Phys. Rev. C, Vol. 44, No. 1, pp. R1745.

[12] J.E. Draper et al, 1990, "Spins in superdeformed bands in the mass 190 region", Phys. Rev. C, Vol. 42, No. 5, pp. R1791.

[13] F. Xu, J. Hu, 1994, "Cranking Bohr-Mottelson Hamiltonian applied to superdeformed bands in A~190 region", Phys. Rev. C, Vol. 49, No. 5, pp. 1449-1453.

[14] A.M. Khalaf, M.M. Sirag and M.M. Taha, 2013, "Spin assignment and behavior of superdeformed bands in A~ 150 mass region", Turkish J. Phys., Vol. 37, No. 1, pp. 49-62.

[15] S. Flibotte, et al, 1993, "ΔI=4 bifurcation in a superdeformed band: Evidence for a C4 symmetry", Phys. Rev. Lett., Vol. 71, No. 26, pp. 4299-4302.

[16] B. Cederwall, et al, 1994, "New features of superdeformed bands in ^{194}Hg", Phys. Rev. Lett., Vol. 72, No. 20, pp. 3150-3153.

[17] A. J. Semple et al, 1996, "Energy Staggering in Superdeformed Bands in ^{131}Ce, ^{132}Ce, and ^{133}Ce", Phys. Rev. Lett., Vol. 76, No. 20, pp. 3671-3674.

[18] A.M. Khalaf, M.D. Okasha and E. Ragheb, 2016, " Anomalous ΔI = 2 Transition Energy Staggering In Nuclear Superdeformed Bands ", Austr. J. Phys. Basic and Appl. Sci., Vol. 10, No. 11, pp. 192-199.

[19] M.P. Carpenter et al, 1995, " Identification of the unfavored N=7 superdeformed band in ^{191}Hg", Phys. Rev. C, Vol. 51, No. 5, pp. 2400-2405.

[20] L.P. Farris et al, 1995, "Neutron blocking and delayed proton pair alignment in

superdeformed ^{195}Pb", Phys. Rev. C, Vol. 51, No. 5, pp. R2288.

[21] L.M. Hilbert, et al, 1996, " Superdeformed structures in 197,198Pb ", Phys. Rev. C, Vol. 54, No. 5, pp. 2253-2258.

[22] G. Hackman *et al*, 1997, " Structure of superdeformed bands in 195Hg ", Phys. Rev. C, Vol. 55, No. 1, pp. 148-153.

[23] A.M.Khalaf M. Kotb and T.M. Awwad, 2017, "Nature of cross-talk transitions and ΔI=1 energy staggering in signature partners of odd mass superdeformed nuclei", Int .J. Mod. Phys. Rev. E, Vol. 26, No. 4, pp. 1750011.

[24] C.S. Wu and Zhau, 1997, " ΔI=1 staggering in odd superdeformed nuclei", Phys. Rev. C, Vol. 56, No. 4, pp. 1814-1820.

[25] A.M. Khalaf et al, 2020, "APPEARANCE OF Delta I= 1 STAGGERING EFFECTS IN SIGNATURE PARTNERS OF ODD SUPERDEFORMED TI AND Pb NUCLEI", University of Bucharest. Scientific Bulletin Series A, Vol. 82, No. 3, pp. 231-244.

[26] A.M. Khalaf et al, 2020, "Description of Superdeformed Bands of the Isotones N=113 for Nuclear Mass Region A~190", Phys. Atomic Nucl., Vol. 83, No. 3, pp. 866-878.

[27] A.A. Abdalaty et al, 2020, "Extended Exponential Model with Pairing Attenuation and Investigation of Energy Staggering and Identical Bands Effects in Superdeformed Thallium Nuclei", Phys. Atomic Nucl., Vol. 83, No. 3, pp. 849-858.

[28] A.M. Khalaf et al, 2019, " Investigation of energy staggering effects in Thorium isotopes in framework of interacting vector Boson model ", Nucl. Phys. A, Vol. 988, pp. 1-8.

[29] T. Byrski et al, 1990, " Observation of identical superdeformed bands in N=86 nuclei", Phys. Rev. Lett., Vol. 64, No. 14, pp. 1650-1653.

[30] C. Baktash, W. Nazarewicz and R. Wyss, 1993, "On the question of spin fitting and quantized alignment in rotational bands", Nucl. Phys. A, Vol. 555, No. 2, pp. 375-407.

[31] A.M. Khalaf, M.M. Sirag and K.M. Abdelmageed, 2016, "Properties of superdeformed bands in A ~ 190 region within the framework of three parameters nuclear softness model", Chinese J. Phys., Vol. 54, No. 3, pp. 329-337.

[32] J.Y. Zeng et al, 2002, "Blocking effect and the moments of inertia of high-K multiquasiparticle bands in ^{172,173,174,175}Hf ", Phys. Rev. C, Vol. 65, No. 4, pp. 044307.

[33] A.M. Khalaf et al, 2020, "Identical bands in doubly even nuclei in framework of variable moment of inertia (VMI) and interacting boson models ", Nucl. Phys. A, Vol. 997, pp. 121719.

[34] A. Bohr and B.R. Moltelson, 1975, " Nuclear Structure v 2", Benjamin, New York.

[35] M. A. J. MARISCOTTI, G. SCHARFF-GOLDHABER, and B. BUCK, 1969, " Phenomenological Analysis of Ground-State Bands in Even-Even Nuclei", Phys. Rev., Vol. 178, No. 4, pp. 1864-1886.

[36] W. Nazarewicz et al, 1990, "Natural-parity states in superdeformed bands and pseudo SU(3) symmetry at extreme conditions", Phys. Rev. Lett., Vol. 64, No. 14, pp. 1654-1657.

[37] C. Rigollet, P. Bonche and H. Flocard, 1999, " Microscopic study of the properties of identical bands in the A=150 mass region ", Phys. Rev. C, Vol. 59, No. 6, pp. 3120-3127.

[38] X.T. He et al, 2005, "The i13/2 proton intruder orbital and the identical superdeformed bands in $^{193, 194, 195}$Tl", Eur. Phys. J. A, Vol. 23, No. 2, pp. 217-222.

[39] Y. J. Chen et al, 2005, "Theoretical simulation for identical bands", Eur. Phys. J. A, Vol. 24, No. 2, pp. 185-191.

[40] A.M. Khalaf, K.E. Abdelmageed and M.M. Sirag, 2015, "Description of the yrast superdeformed bands in even-even nuclei in A ~ 190 region using the nuclear softness model", Turck. J. Phys., Vol. 39, No. 2, pp. 178-186.

[41] F.S. Stephens et al, 1990, " Spin alignment in superdeformed Hg nuclei ", Phys. Rev. Lett., Vol. 64, No. 22, pp. 2623-2626.

[42] F.S. Stephens, 1990, " Spin alignment in superdeformed rotational bands ", Nucl. Phys. A, Vol. 520, pp. c91-c104.

Design and Simulation of Terahertz Apodized Bragg Grating Using Coplanar Stripline Transmission Line a 1 μm-thin Membrane

Walid Gomaa[1] **and Thomas Darcie**[2]

[1] Department of Engineering Physics, Military Technical College, Cairo, Egypt
[2] Department of Electrical and Computer Engineering, University of Victoria, Victoria, Canada

E-mail: `walid.gomaa@mtc.edu.eg`

Abstract.
A guided-wave THz System-on-Chip THz System-on-Chip (TSoC) is emerging as an attractive alternative to the routine free-space THz systems to reduce physical bulk, propagation loss, pulse dispersion and cost of free-space THz systems. Recently, our research group succeeded in demonstrating a novel waveguided TSoC based on the coplanar stripline Coplanar Stripline (CPS) transmission lines on a 1 μm-thin Silicon Nitride membrane. The novelty of this membrane-based platform was bonding the transmitter and receiver directly on the transmission line to eliminate the radiation loss by the routine THz optics. Besides, the delicate thin-membrane dramatically reduces the dielectric loss of the platform which results in low-loss and low-dispersion THz-bandwidth pulses.

In this paper, we introduce the design, theoretical model and the simulation results of a novel THz Apodized Bragg Grating (TABG) based on a periodic reflection of consecutive different impedance sections of CPS transmission lines on the membrane platform. This structure was introduced based on theoretical modelling and the simulation results. The CPS-TABG consists of 20 $\lambda/2$-length cascaded sections. The central frequency of the bandstop filter was selected around 0.8 THz to avoid signal distortion of the received terahertz pulse.

1. Introduction

THz systems have improved in performance and utility, driven by the refinement of Photoconductive Antenna (PCA) and sub-picosecond lasers used to generate and detect THz-bandwidth signals [1, 2, 3]. Many promising applications have emerged in sensing, imaging, spectroscopy, science, inspection, and communication [4, 5, 6, 7, 8, 9, 10, 11, 12, 13, 14], but these PCA systems remain bulky and expensive [15, 16, 17, 18, 19, 20].

One significant area of work targets developing integrated, compact and efficient TSoC technologies in which guided-wave THz systems replace free-space THz systems in use today [21, 22, 23]. A low-loss, low-distortion and low-pulse-dispersion TSoC had been demonstrated using Photoconductive Switch (PCS) devices bonded directly to a CPS transmission line defined using photolithography on 1μm-thin Si_3N_4 membrane [24]. Based on this capability, it is now possible to develop different components for the TSoC including active devices (e.g. photoconductors, photomixers, amplifiers, etc.) and passive devices (e.g. filters, couplers, power dividers, chokes, etc.) [25, 26, 27, 28, 29]

In this work, we introduce a novel TABG integrated into the CPS transmission line as one of the passive components enabled by the membrane-based platform introduced in [24]. This filter is designed using standard microwave analysis techniques to reject specific spectral components from a THz-bandwidth transmitted pulse. The rejection bandwidth of the TABG was selected around $f_c = 0.8$ THz to push the second-order rejection ($f_c = 1.6$ THz) beyond 1.5 THz. The crucial factor is the limitation of the entire length of the filter structure; the design should introduce a deep rejection at the stopband and minimum losses at the passband, while longer length of the structure introduces higher loss. In addition, the geometry of the CPS transmission line (i.e. stripline width and separation) is limited by the fabrication facilities; photolithography can provide a high-resolution stripline with feature sizes easily within 1-3 μm. In this work, the structure was optimized near a more conservative resolution (3-5 μm).

2. Mathematical Model

Optical Fiber Bragg Filters (FBGs) have become ubiquitous in optical communication systems due to low loss and high frequency selectivity. They are fabricated by changing the index of refraction of the fiber core periodically or quasi-periodically. The Bragg condition is a coherent combination of the light reflected from each period producing a large reflection at a particular wavelength [30]. The first-order Bragg condition is defined as[31, 32]:

$$\lambda_B = 2n_{eff}\Lambda \tag{1}$$

where λ_B is the free-space Bragg wavelength, n_{eff} is the effective refractive index of the fiber core and Λ is the periodic length of the grating. This phenomenon and the popularity of the Fiber Bragg Filter (FBG) inspired us to design a Bragg grating on a CPS transmission line by constructing a periodic reflection based on impedance mismatch of consecutive different impedance CPS sections.

The simplest structure of the proposed THz Bragg Filter (TBG) shown in Fig. 1 consists of (N) identical cascaded sections of length $\Lambda = \lambda_B/2\sqrt{\varepsilon_{eff}}$, each section consisted of three parts (2 parts with length $L_\circ = \Lambda/4$ and impedance of $Z_\circ$ and the third part in between with length $L_1 = \Lambda/2$ and impedance of Z_1).

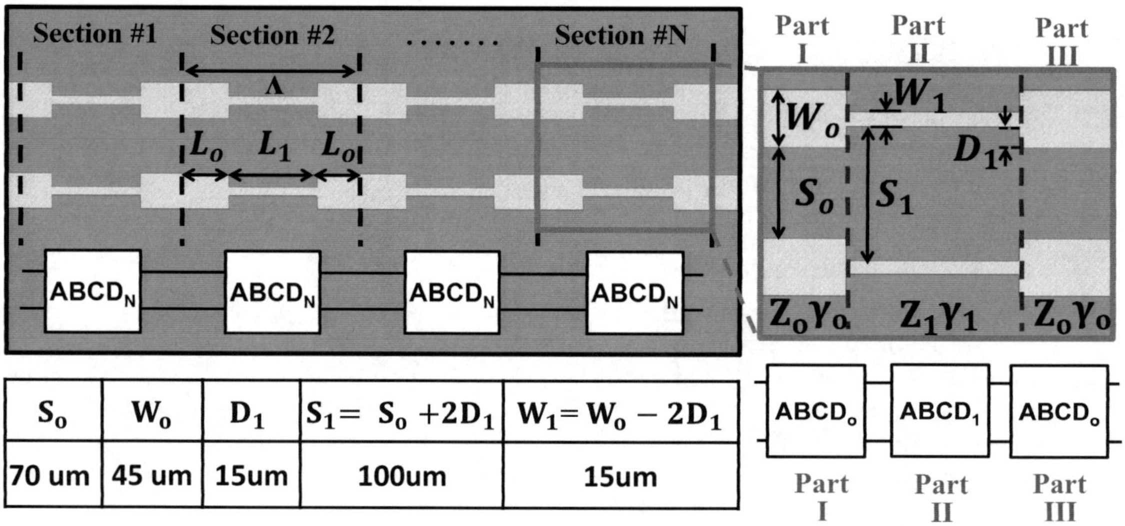

S_o	W_o	D_1	$S_1 = S_o + 2D_1$	$W_1 = W_o - 2D_1$
70 um	45 um	15um	100um	15um

Figure 1. Structure of TBG integrated in CPS transmission line.

We modeled each of these sections using a chain-parameter matrix representation (ABCD transmission matrix) of lossy transmission lines, which relates the terminal voltages and currents of two-port systems [33]:

$$\begin{bmatrix} V_{out} \\ I_{out} \end{bmatrix} = \begin{bmatrix} A & B \\ C & D \end{bmatrix} \begin{bmatrix} V_{in} \\ I_{in} \end{bmatrix} \tag{2a}$$

where:

$$\begin{bmatrix} A & B \\ C & D \end{bmatrix} = \begin{bmatrix} cosh\gamma L & Z sinh\gamma L \\ (1/Z) sinh\gamma L & cosh\gamma L \end{bmatrix} \tag{2b}$$

Where $\gamma = \alpha + i\beta$ is the complex propagation constant, L is the length and Z is the complex impedance of the CPS transmission line. We used a 2D simulation to obtain approximate values of different geometries CPS transmission lines (more details in Sec. 3.1)

The ABCD transmission matrix of each section is modelled as a three cascaded ABCD matrices (two parts with length $L_\circ = \Lambda/4$ and impedance of $Z_\circ$ and the third part (in-between) with length $L_1 = \Lambda/2$ and impedance of Z_1):

$$\begin{bmatrix} A_S & B_S \\ C_S & D_S \end{bmatrix} = \begin{bmatrix} A_\circ & B_\circ \\ C_\circ & D_\circ \end{bmatrix} \begin{bmatrix} A_1 & B_1 \\ C_1 & D_1 \end{bmatrix} \begin{bmatrix} A_\circ & B_\circ \\ C_\circ & D_\circ \end{bmatrix} \tag{3}$$

For N identical sections of the TBG, the ABCD transmission matrix is :

$$\begin{bmatrix} A_N & B_N \\ C_N & D_N \end{bmatrix} = \begin{bmatrix} A_S & B_S \\ C_S & D_S \end{bmatrix}^N \tag{4}$$

The S-parameters of the TBG were calculated from the ABCD matrix using the following equations [34]:

$$S_{11} = \frac{A_N Z_{\circ 2} + B_N - C_N Z_{\circ 1}^* Z_{\circ 2} - D_N Z_{\circ 1}^*}{A_N Z_{\circ 2} + B_N + C_N Z_{\circ 1} Z_{\circ 2} + D_N Z_{\circ 1}} \tag{5a}$$

$$S_{21} = \frac{2(R_{01} R_{02})^{1/2}}{A_N Z_{\circ 2} + B_N - C_N Z_{\circ 1} Z_{\circ 2} - D_N Z_{\circ 1}} \tag{5b}$$

where $Z_{\circ 1}$ is the complex source impedance and $Z_{\circ 2}$ is the complex load impedance. It should be noted that the source and load impedances are the impedance of the CPS transmission line $Z_\circ$ (i.e. $Z_{\circ 1} = Z_{\circ 2} = Z_\circ$).

Matlab was used to calculate and plot the S-parameters of the TBG for different parameters. Figure 2(a) illustrates that increasing N provides a lower level of ripple, and deeper rejection of S_{21} (Fig. 2(b)). The length of the TBG restricts increasing the number of sections because a longer structure introduces more attenuation.

The depth of the cut-off D is one of the most important parameters affecting the design performance. Deeper depth means narrower CPS lines with wider separation between lines which leads to higher impedance, consequently higher impedance mismatch and reflection. Increasing D provides a deeper stopband width of S_{21}, but it increases the reflection ripple of sidelobe level of the passband width of S_{11} (as shown in Fig. 3).

To improve the sidelobe performance, we used an apodization profile similar to that formerly used in FBGs to reduce the side lobes of the reflection spectrum [35, 36, 37, 38, 39]. The cut-off depth D of the first four sections was gradually increased from both sides of the TABG structure as shown in Fig. 4.

The ABCD matrix of the TABG can be expressed as:

Journal of Physics: Conference Series **2304** (2022) 012015 doi:10.1088/1742-6596/2304/1/012015

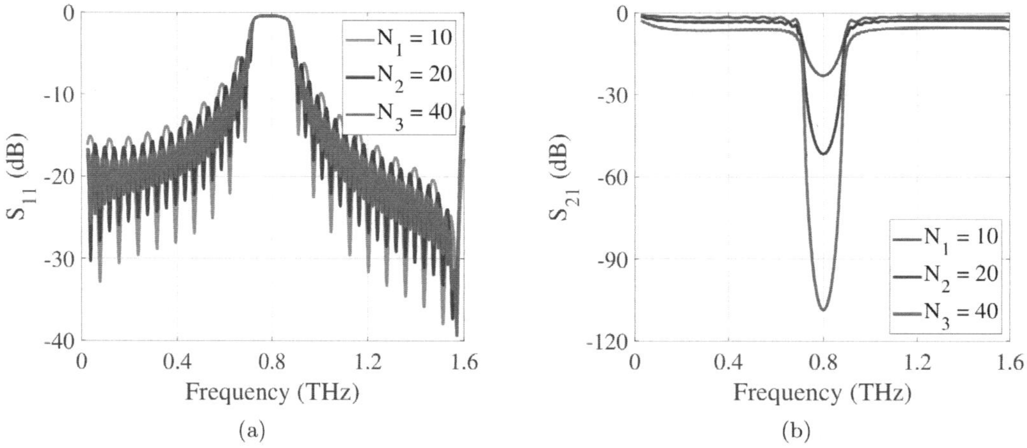

Figure 2. (a) Reflection coefficient of TBG for different number of sections, N. (b) Transmission coefficient of TBG for different N.

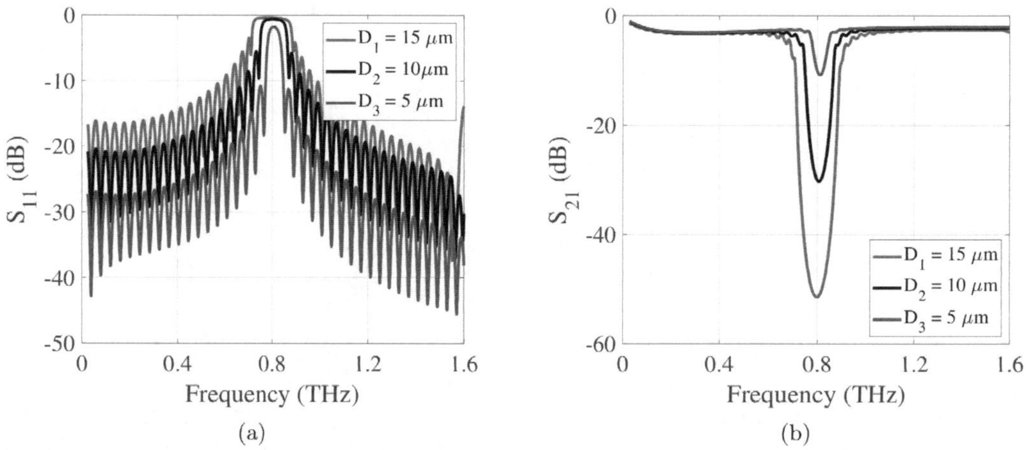

Figure 3. (a) Reflection coefficient of TBG for different cut-off depth, D. (b) Transmission coefficient of TBG for different D.

$$\begin{bmatrix} A_T \, B_T \\ C_T \, D_T \end{bmatrix} = \begin{bmatrix} A_1 \, B_1 \\ C_1 \, D_1 \end{bmatrix} \times \begin{bmatrix} A_2 \, B_2 \\ C_2 \, D_2 \end{bmatrix} \times \begin{bmatrix} A_3 \, B_3 \\ C_3 \, D_3 \end{bmatrix} \times \begin{bmatrix} A_4 \, B_4 \\ C_4 \, D_4 \end{bmatrix} \times \begin{bmatrix} A_U \, B_U \\ C_U \, D_U \end{bmatrix}^{N-8} \times \begin{bmatrix} A_4 \, B_4 \\ C_4 \, D_4 \end{bmatrix} \times \begin{bmatrix} A_3 \, B_3 \\ C_3 \, D_3 \end{bmatrix} \times \begin{bmatrix} A_2 \, B_2 \\ C_2 \, D_2 \end{bmatrix} \times \begin{bmatrix} A_1 \, B_1 \\ C_1 \, D_1 \end{bmatrix}. \quad (6)$$

where the ABCD matrix of each section can be calculated using Eqn. 2b according to the dimensions of each section (mentioned in Fig. 4). The effect of the apodization profile is clear in Fig. 5, where the sidelobe levels are significantly decreased, while the depth of the stopband is acceptable.

Figure 6 illustrates that the bandwidth of the stopband is proportional to the central frequency, and the structure with higher central-frequency provides a lower bandpass loss (i.e. due to the shorter length) while it introduces a wider stopband bandwidth.

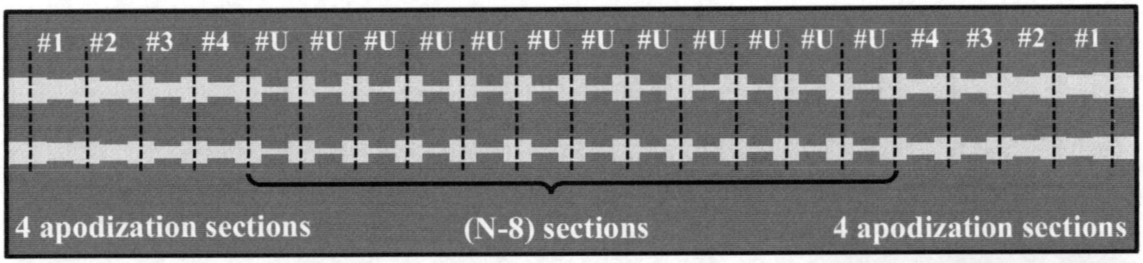

Section #	#1	#2	#3	#4	#U
D_N	5um	7.5um	10um	12.5um	15um
$W_N = W_o - 2D_N$	35um	30um	25um	20um	15um
$S_N = S_o + 2D_N$	80um	85um	90um	95um	100um

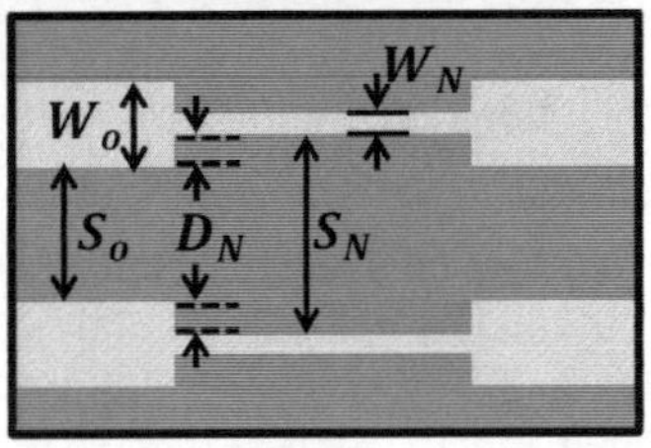

Figure 4. Structure of the TABG integrated inCPS transmission line.

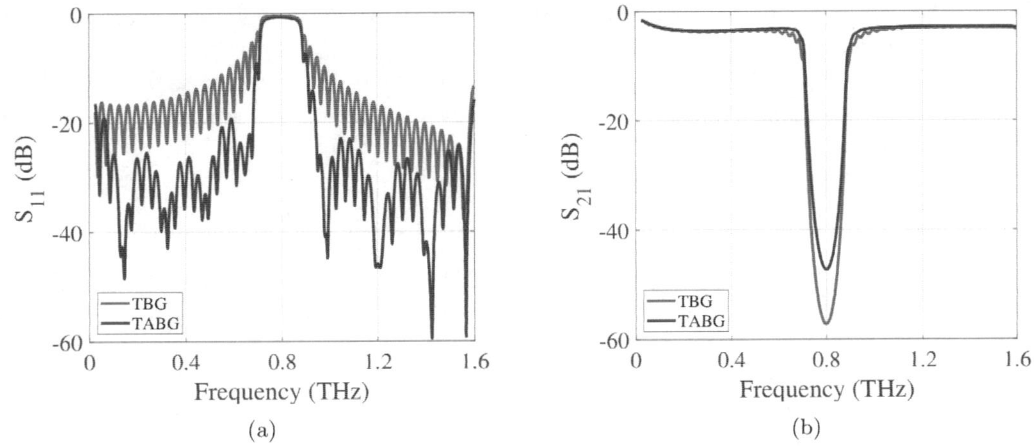

Figure 5. The role of apodization profile on the S-parameters of TBG: (a) Reflection coefficient S_{11}. (b) Transmission coefficient S_{21}.

3. Simulation Results

The design was simulated in both time and frequency domain using ANSYS HFSS software [40]. Frequency-domain simulations are used to gain an accurate and deep insight of the TABG on a CPS and optimize the parameters and dimensions of the structure. Time-domain simulations are used to characterize pulse excitation of the design and evaluate the spectrum of the received pulses after propagating through the design structure. Long simulation times are required to simulate the design in its entirety due to the extreme dimensional ratio of thickness to the length of the membrane (1:10000). The following material parameters are used in the simulations; For silicon nitride: $\epsilon_r = 7.6$, $\mu_r = 1$, and $tan\delta_e = 0.00526$ [41]. For the gold: $\sigma_{au} = 4.1 \times 10^7$S/m with negligible surface roughness.

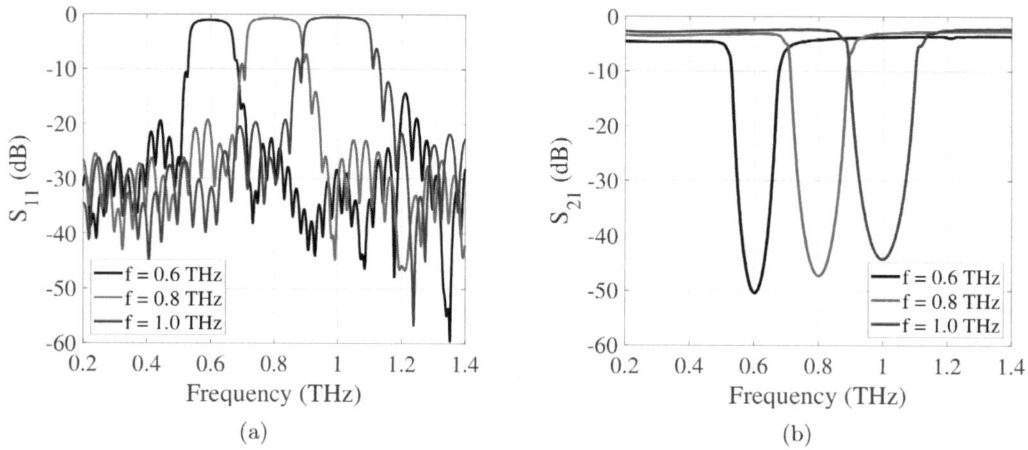

Figure 6. S-parameters of TABG for different central frequency: (a) Reflection coefficient. (b) Transmission coefficient.

3.1. Frequency Domain Simulations

First, the ANSYS High Frequency Structure Simulator (HFSS) frequency-domain solver was used to characterize CPS transmission-line complex impedance, Z, and complex propagation constant $\gamma = \alpha + i\beta$. Figure. 7 shows that separation between CPS transmission line, S, and stripline width, W, have significant impact on Z and γ. The optimum dimensions of CPS which provide minimum attenuation were found to be $W_\circ = 45\ \mu$m and $S_\circ = 70\ \mu$m. Moreover, increasing the depth of step size (i.e. wider S_N and narrower W_N) provides higher impedance and consequently higher reflection.

Then, the ANSYS HFSS frequency-domain solver was used to optimize the structure of the TABG on a CPS transmission-line shown in Fig. 4. The reflection coefficient S_{11} and the transmission coefficient S_{21} obtained by HFSS (shown in Fig. 8) are consistent with Matlab results. The results obtained by HFSS show a deeper stopband and lower ripple level than that obtained by the Matlab model due to the approximations mentioned in Eqn.5. The difference between the transmission coefficients of the Perfect Electric Conductor (PEC) and gold provides an estimate of the structure's conductor loss.

Figure 9 shows the propagation of the electric field through the TABG for different frequencies in passband and stopband. The propagation of the electric field at passband (i.e. 0.8 THz and 1.6 THz) has not been affected through the TABG structure except the radiation at the higher frequency (1.6 THz). The electric field at 0.8 THz (i.e. at stopband) cannot propagate to the output port and the reflection from TABG produces a standing wave.

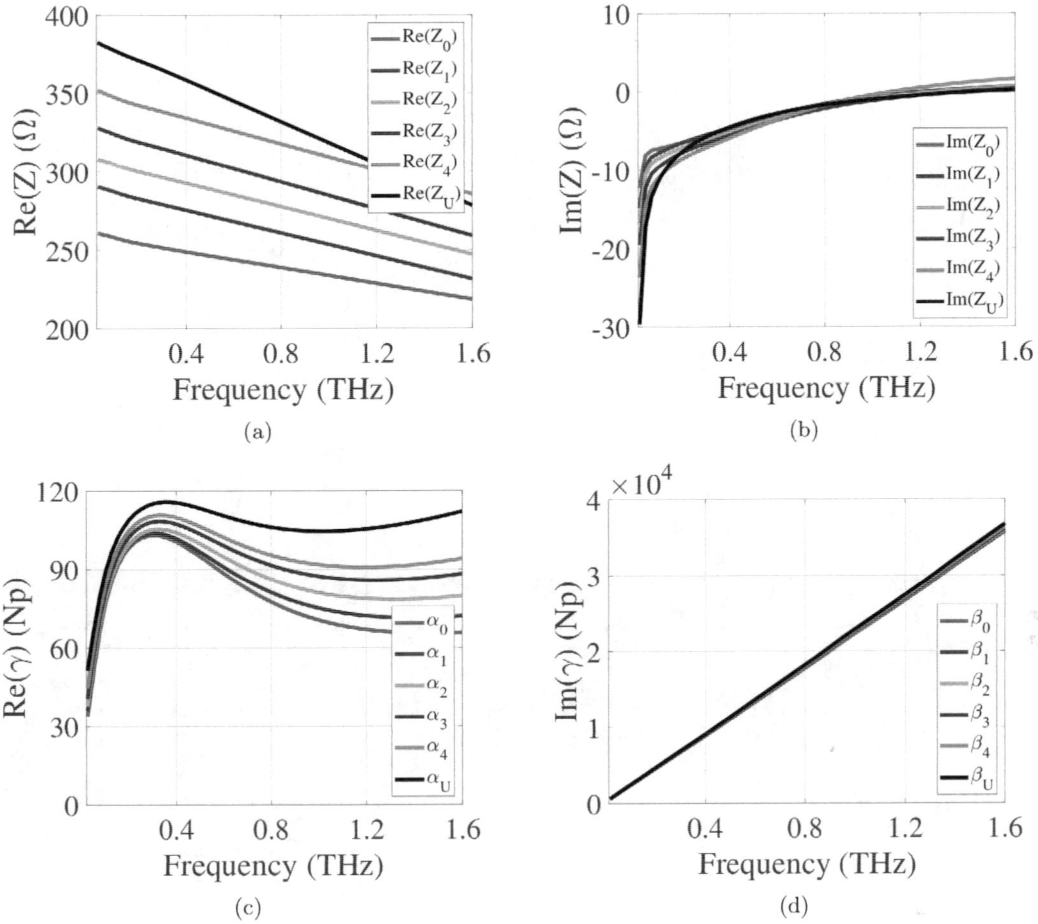

Figure 7. Frequency domain simulations results for CPS transmission line on thin membrane. (a) $Re(Z)$. (b) $Im(Z)$. (c) $Re(\gamma) = \alpha$. (d) $Im(\gamma) = \beta$.

3.2. Time Domain Simulations

The time-domain simulation was used to simulate the electric pulse generated from PCS and evaluate the pulse delivered at the output port (temporal and spectrum).

Figure 10 illustrates the entire structure which consists of the TABG integrated onto a CPS transmission line, the tapering structure and the Bias-T configuration.

The overall dimensions of the Low Temperature Grown-Gallium Arsenide (LTG-GaAs) devices used for transmitter and receiver are 20 μ m $\times$ 40 μm which restricts the dimensions of CPS transmission line at the input and output ports to be W_i = 10 μm and S_i = 10 μm, which provides a high attenuation. Therefore,, we designed a CPS taper from (W_i = 10 μm, S_i = 10 μm) to ($W_\circ$ = 45 μm, $S_\circ$ = 70 μm) to reduce this attenuation. Moreover, this tapering provides an additional degree of flexibility in customizing different CPS characteristics. For example, wider striplines enables us to change the cutoff depth applied to the apodization sections at the beginning and end of the CPS transmission line. The Bias-T configuration consists of an interdigitated capacitor, C, and meander inductor, L. The interdigitated capacitor blocks the DC bias from reaching the receiving PCS while the inductor passes the DC bias voltage to the PCS for the transmitting PCS without leakage of the THz pulse.

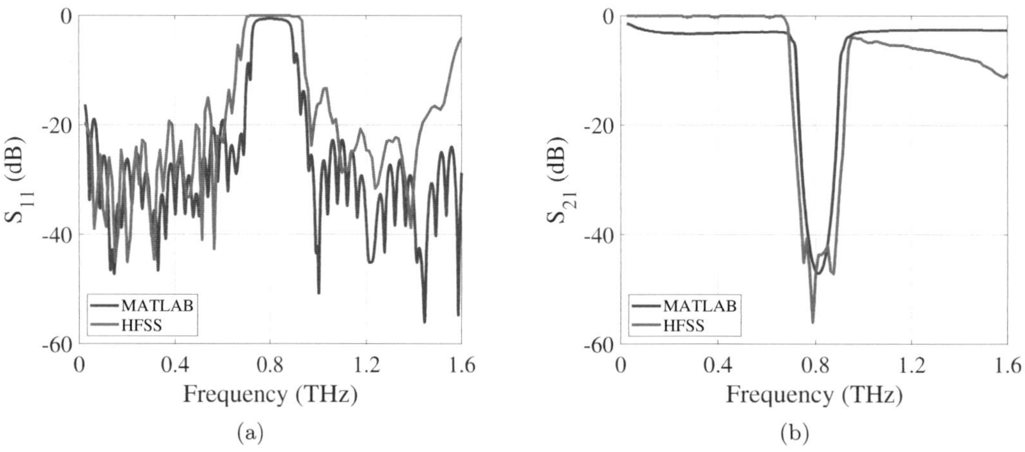

Figure 8. Frequency domain simulations results for TABG on thin membrane. (a) Reflection coefficient S_{11}. (b) Transmission coefficient S_{21}.

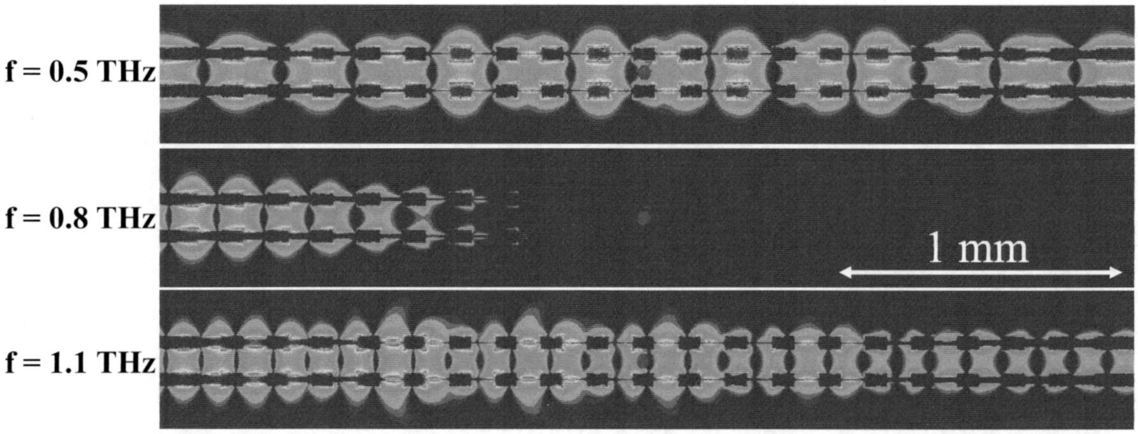

Figure 9. Electric field propagation continuous wave through TABG at different frequency: (a) 0.5 THz, (b) 0.8 THz and (c) 1.1 THz.

Figure 11(a) shows the signal at the input port (i.e. the reflection from the TABG structure) and the picosecond-pulse at the output port. The reflected signal represents a beat of the rejection bandwidth spectrum. Figure 11(b) illustrates the Fast Fourier Transform (FFT) of the received pulse. The rejection bandwidth of TABG is obvious around f_c = 0.8 THz. It is clear that the spectral power around the second-order rejection band (f_c = 1.6 THz) is minor and does not have a large impact on the received pulse.

4. Conclusion

We demonstrate the design and simulation results of a THz apodized Bragg grating based on periodic reflection of consecutive different impedance sections of coplanar stripline (CPS) transmission lines. The structure consists of 20 $\lambda/2$-length cascaded sections. The rejection bandwidth of the TABG was selected around f_c = 0.8 THz to push the second-order rejection ($f_{c2} = 2f_c$ =1.6 THz) beyond 1.5 THz.

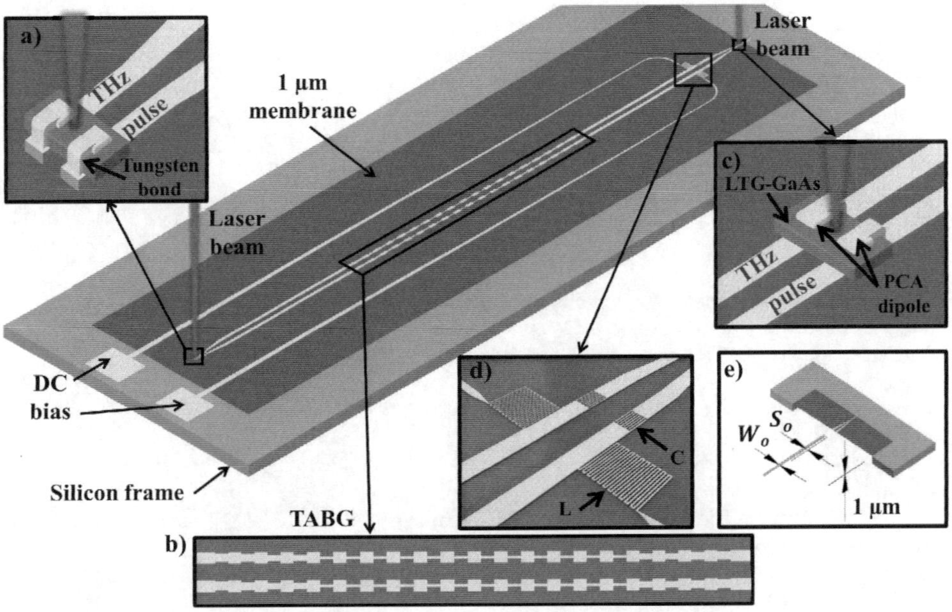

Figure 10. Structure of TABG integrated in CPS transmission line on Si_3N_4 membrane. (a) Rendering of transmitting PCS. (b) Rendering of TABG structure. (c) Rendering of receiving PCS. (d) Rendering of bias-T configuration. (e) Cross-section of structure.

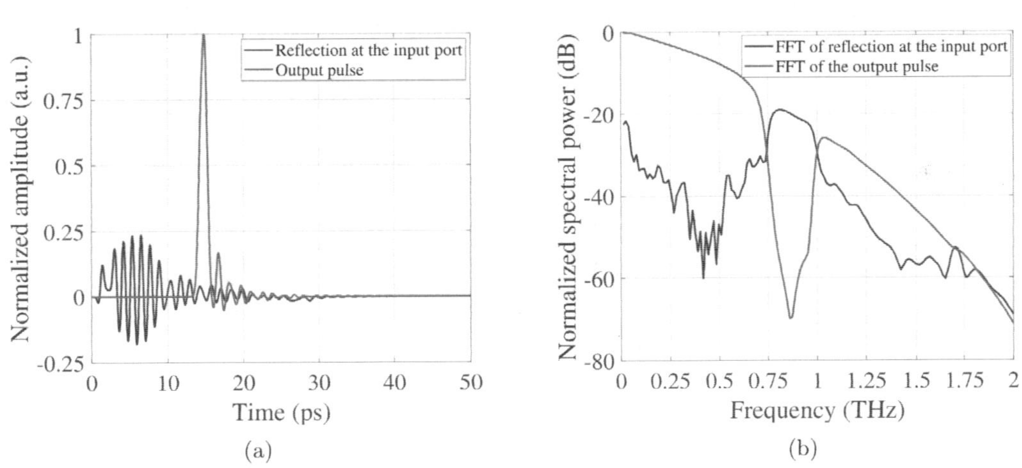

Figure 11. Time domain simulations results for TABG on the thin membrane: (a) Reflection coefficient S_{11}. (b) Transmission coefficient S_{21}.

References

[1] D. Auston, 1983, "Impulse response of photoconductors in transmission lines," IEEE Journal of Quantum Electronics, vol. 19, no. 4, pp. 639–648.

[2] D. H. Auston et al, 1984, "Picosecond photoconducting hertzian dipoles," Applied Physics Letters, vol. 45, no. 3, pp. 284–286.

[3] M. B. Ketchen et al, 1986, "Generation of subpicosecond electrical pulses on coplanar transmission lines," Applied Physics Letters, vol. 48, no. 12, pp. 751–753.

[4] A. G. Davies et al, 2002, "The development of terahertz sources and their applications," Physics in Medicine and Biology, vol. 47, no. 21, pp. 3679–3689.

[5] M. C. Kemp, 2006, "Millimetre wave and terahertz technology for the detection of concealed threats: a review," in Optics and Photonics for Counterterrorism and Crime Fighting II, vol. 6402, International Society for Optics and Photonics. SPIE, pp. 90–108.
[6] M. Tonouchi, 2007, "Cutting-edge terahertz technology," Nature Photonics, vol. 1, no. 2, pp. 97–105.
[7] C. Baker et al, 2007, "Detection of concealed explosives at a distance using terahertz technology," Proceedings of the IEEE, vol. 95, no. 8, pp. 1559–1565.
[8] M. Neshat et al, 2007, "An integrated continuous-wave terahertz biosensor," in Terahertz for Military and Security Applications V, vol. 6549, International Society for Optics and Photonics SPIE, pp. 113–120.
[9] K. Fukunaga et al, 2010 "Terahertz spectroscopy applied to the analysis of artists' materials," Applied Physics A, vol. 100, p. 591–597.
[10] V. Sharma et al, 2011, "Terahertz technology and its applications," in 5th International Conference on Advanced Computing Communication Technologies, pp. 175–178.
[11] A. Y. Pawar et al, 2013, "Terahertz technology and its applications," Drug Invention Today, vol. 5, no. 2, pp. 157–163.
[12] J. H. Son et al, 2019, "Potential clinical applications of terahertz radiation," Journal of Applied Physics, vol. 125, no. 19, p. 190901.
[13] L. Yu et al, 2019, "The medical application of terahertz technology in non-invasive detection of cells and tissues: opportunities and challenges," RSC Adv., vol. 9, pp. 9354–9363.
[14] R. Zhou et al, 2019, "Biological applications of terahertz technology based on nanomaterials and nanostructures," Nanoscale, vol. 11, pp. 3445–3457.
[15] K. Sengupta et al, 2018, "Terahertz integrated electronic and hybrid electronic/photonic systems," Nature Electronics, vol. 1, no. 12, pp. 622–635.
[16] K. Sengupta eta al, 2017, "THz silicon systems on chip: EM-circuits-systems codesign approach," 42nd International Conference on Infrared, Millimeter, and Terahertz Waves (IRMMW-THz). IEEE.
[17] L. Dazhang et al, 2009, "On-chip terahertz Goubau line waveguides with integrated photoconductive emitters and mode discriminating detectors," Applied Physics Letters, vol. 95, no. 9, p. 092903.
[18] T. Nagatsuma et al, 2016, "Advances in terahertz communications accelerated by photonics," Nature Photonics, vol. 10, no. 6, pp. 371–379.
[19] S. Koenig et al, 2013, "Wireless sub-THz communication system with high data rate," Nature Photonics, vol. 7, no. 12, pp. 977–981.
[20] L. A. Samoska, 2011, "An overview of solid-state integrated circuit amplifiers in the submillimeter-wave and THz regime," IEEE Transactions on Terahertz Science and Technology, vol. 1, no. 1, pp. 9–24.
[21] B. Vidal et al, 2012, "Photonic technologies for millimeter- and submillimeter-wave signals," Advances in Optical Technologies, vol. 2012, pp. 1–17.
[22] L. Smith et al, 2014, "THz field enhancement by antenna coupling to a tapered thick slot waveguide," J. Lightwave Technol., vol. 32, no. 20, pp. 3676–3682.
[23] R. Smith etal, 2017, "Photoconductive generation and detection of THzbandwidth pulses using near-field coupling to a free-space metallic slit waveguide," Optics Express, vol. 25, no. 22, p. 26492.
[24] R. Smith et al, 2019, "Demonstration of a low-distortion terahertz system-on-chip using a CPS waveguide on a thin membrane substrate," Optics Express, vol. 27, no. 10, p. 13653.
[25] W. Gomaa t al, 2020, "Terahertz low-pass filter based on cascaded resonators formed by cps bending on a thin membrane," Opt. Express, vol. 28, no. 21, pp. 31967–31978.
[26] W. Gomaa et al, 2020, "Terahertz power divider using symmetric CPS transmission line on a thin membrane," IEEE Access, vol. 8, pp. 214425–214433.
[27] L. Smith et al, 2021, "Tapered transmission lines for terahertz systems," Opt. Express, vol. 29, no. 11, pp. 17 295–17 303.
[28] L. Smith et al, 2021, "Characterization of a split-ring-resonator-loaded transmission line at terahertz frequencies," Opt. Express, vol. 29, no. 15, pp. 23282–23289.
[29] W. Gomaa, 2021, "Terahertz System-on-Chip using coplanar stripline transmission line on thin membrane," https://dspace.library.uvic.ca/handle/1828/12504.
[30] R. Kashyap, 2009, "Fiber Bragg Gratings," Academic Press.
[31] Z. Yu, 2008, "Fibre Bragg grating components for filtering, switching and lasing," Ph.D. dissertation, KTH information and communication technlogy.
[32] A. Othonos et al, 1999, "Fiber Bragg gratings : fundamentals and applications in telecommunications and sensing", Artech House.
[33] C. R. Paul, 2007, "Analysis of Multiconductor Transmission Lines," 2nd ed. Wiley-IEEE Press.
[34] D. A. Frickey, 1994, "Conversions between S, Z, Y, H, ABCD, and T parameters which are valid for complex source and load impedances," IEEE Transactions on Microwave Theory and Techniques, vol. 42, no. 2, pp. 205–211.

[35] A. Othonos, 1997, “Fiber Bragg gratings,” Review of Scientific Instruments, vol. 68, no. 12, pp. 4309–4341.
[36] N. H. Sun et al, 2009, “Numerical analysis of apodized fiber Bragg grating using coupled mode theory,” Progress In Electromagnetics Research, vol. 99, pp. 289–306.
[37] E. Gemzicky et al, 2009, “Numerical analysis of reflection characteristics of cascaded non-uniform fiber Bragg gratings,” in Optical Sensors 2009, vol. 7356, International Society for Optics and Photonics. SPIE, pp. 420–428.
[38] O. Ozolins, 2013, “Fiber Bragg grating technology for new generation optical access systems,” in Current Trends in Short- and Long-period Fiber Gratings. InTech.
[39] I. Ashry, et al, 2014, “Investigating the performance of apodized fiber Bragg gratings for sensing applications,” in Proceedings of the 2014 Zone 1 Conference of the American Society for Engineering Education. IEEE.
[40] ANSYS HFSS simulation suite, Release 2020 R1.
[41] G. Cataldo, et al, 2012, “Infrared dielectric properties of low-stress silicon nitride,” Optics Letters, vol. 37, no. 20, p. 4200.

Study the Optimal Operative Conditions of New Glow Discharge Ion Source for Polymeric Treatment

M.G. Hassan[1], B.A. Soliman[1], A.A. El-Saftawy[1], M.M. Abdelrahman[1], K.E. Abdelmageed[2] and M.A. Hassouba[2]

[1] Accelerators and Ion Sources Department, Nuclear Research Center, Egyptian Atomic Energy Authority, P.O. 13759, Cairo, Egypt
[2] Physics Department, Faculty of Sciences, Benha University, Benha, Egypt

Email: mg819521@gmail.com

Abstract. The design, manufacturing, and systematic parametric evaluation of a modified glow discharge ion source (with a pear-shaped anode and a plane cathode) are detailed to extract low-power, high plasma density, and high-intensity ion beam current for material processing. The study includes determining the optimal operative conditions for ion source performance and optimizing its properties. The ion beam current was measured at various argon gas pressures and anode-cathode distances as well as its measurements at various ion source - collector plate distances. At the optimum settings, the influence of discharge pressure on the output current was evaluated using various gases (argon, oxygen, nitrogen and hydrogen). In addition, the influence of applied negative voltages on the extractor electrode (different hole diameters) on the generated ion beam current is being explored. Finally, SIMION computer program evaluated and studied ion beam optics. The simulation and experimental results for a highly efficient ion beam derived from the planned ion source agreed well.

Keywords: Pear-shaped anode, Plane cold cathode, glow discharge ion source.

1. Introduction

Ion source is a device for obtaining direct flow of ions. Ion sources are used in charged particle accelerators for thin film technology, mass separators, plasma accelerators, and a variety of other applications. The primary objective of this study is to design and improve of a home- made direct current glow discharge ion source for material processing. Specifications for this ion source are required to fit the proposed uses, including the generation of hundreds of microampere beams for industrial polymerization. In this glow discharge ion sources, the pear-shaped anode is connected to positive potential ranging from 500 to 5000 V, while the cold - plane cathode is earthed .The ionization takes place when the primary electrons oscillates in between the pear-anode and the plane-cold cathode and collides with gas molecules inside the source . The variation in potential between the cathode and the anode creates a field gradient that acts to draw ions from the plasma and through the hole within the cathode plate. Copper is used for the pear-shaped anode due to its high ionization coefficient [1]. Simultaneously, the cold plane cathode is constructed of aluminum with a high secondary electron emission coefficient, thereby increasing plasma density and ion beam current [2]. The generated ion beam current can be enhanced and confined by applying negative extraction voltages to varying hole diameters in the extractor electrode. The features of the output ion beam current have been examined utilizing SIMION 3D version 7 computer code. The modelling procedure was utilized to evaluate the extraction properties in order to produce an ion beam with low beam emittance and high brightness while being carried to the appropriate target .Our home-made ion source is distinguished by its simple design, long operation without filament, lack of maintenance as well as low pressure operation . It could be used for a wide range of purposes, including Sputtering and materials surface modification. The optimal operative condition for this ion source was found to be governed by two forces: the first is the anode's mechanical constriction (pear-anode shape), and the second is the electrical confinement caused by the potential difference between the pear-anode and cold plane cathode. These parameters lead to stability of operation, high density plasma and low gas consumption, consequently; a maximum output ion current could be obtained.

2. Experimental Details

Fig. 1 depicts a schematic diagram and photographic images of a newly constructed glow discharge ion source and its electrical circuit. It is composed of a non-magnetic cupper anode (pear-shaped), 45 mm diameter and 35 mm length as well as a 6 mm ion exit aperture. The anode's upper side is shaped like a hemisphere to create a concave plasma boundary, which results in high-density focused plasma. The lower side features a Pierce configuration for axial mechanical constriction and electrical discharge confinement, allowing for a more significant output ion beam current [3]. The cathode is an aluminum plane with a 2 mm orifice. The source is assembled inside cylindrical insulator to provide isolation of 50kv. The working gas was fed through a 2 mm diameter orifice in the anode's outer surface, with the gas flow rate adjustable using a finely controlled needle valve. The collector plate that is mounted at various distances from the ion outlet aperture and used to measure the output ion beam current. For initiating the discharge, a digital positive power supply with a voltage range of 0 to 30 kV was employed. The discharge current I_d was measured using a milli-ampere metre with a range of 0 to 10 mA. While the output ion beams current I_b was measured using a micro-ampere meter.

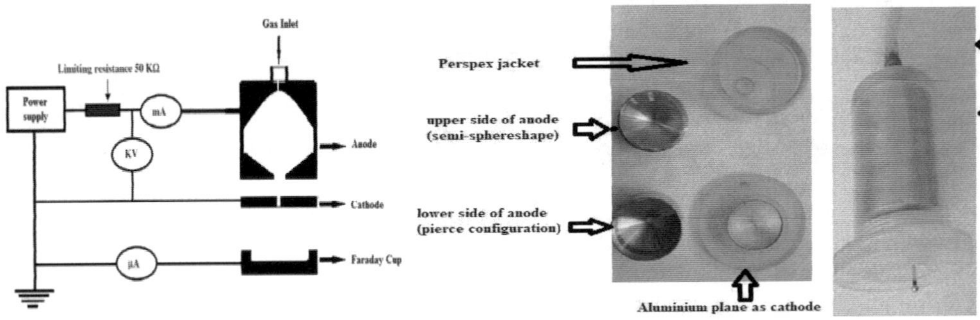

Fig.1 Designed home-made glow discharge ion source and its electrical circuit.

The ion source vessel was purged using a vacuum system. It comprises of a 225 L/ S stainless silicon diffusion pump with an electrical heater and a 15 m^3 / h. rotary fore vacuum pump. To prevent silicon vapors from reaching the ion source vessel, the diffusion pump is encircled by piping system for chilling down to 3 c° during operation using an Industrial chillier of type (cw-5300).

3. Experimental investigations and Results

Fig.2(a-d) shows the relationship among both discharge current, I_d and the discharge voltage, V_d with various Ar^+ gas pressures and varying separated lengths between the anode and the cathode : (a) 1mm , (b) 2mm , (c) 3mm , and (d) 4mm , respectively . It was observed that, the elevation in discharge voltage was found to be paralleled by an improvement in discharge current as shown by this figure which made this kind of ion source suitable for abnormal glow discharge applications.

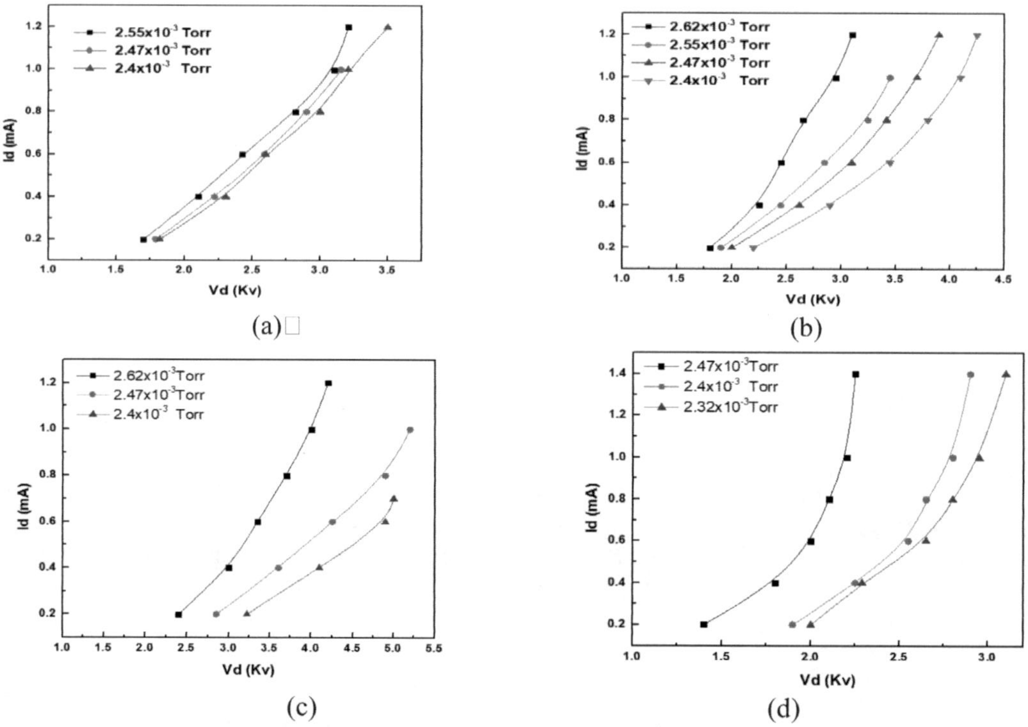

Fig. 2 Discharge voltage and discharge current at various Ar^+ gas pressures and varying distances between the anode and the cathode: (a) 1 mm , (b) 2 mm , (c) 3 mm ,(d) 4 mm

Fig. 3(a-d) demonstrates the relation between the output ion beam current, I_b, and the discharge current, I_d, for various anode - cathode distances: (a) 1 mm, (b) 2 mm, (c) 3 mm, and (d) 4 mm, respectively at various Ar^+ gas pressures. It was noticed that, whenever the discharge current increases, so does the output ion beam current, as previously stated. Also, the optimum length between both the cathode and the anode where found to be at 3mm where a highest possible output ion beam current is 55 μA at $P_{Ar}^+ = 2.4\times10^{-3}$ Torr, and 0.6 mA discharge current, due to a stable high density plasma at this specific distance (see fig.1). Beyond this optimal value, plasma instability and mechanical design problems appeared which causes breakdown [4]. Under these conditions, the plasma meniscus should to be concave shape and this result in a focused ion beam (minimum beam diameter).

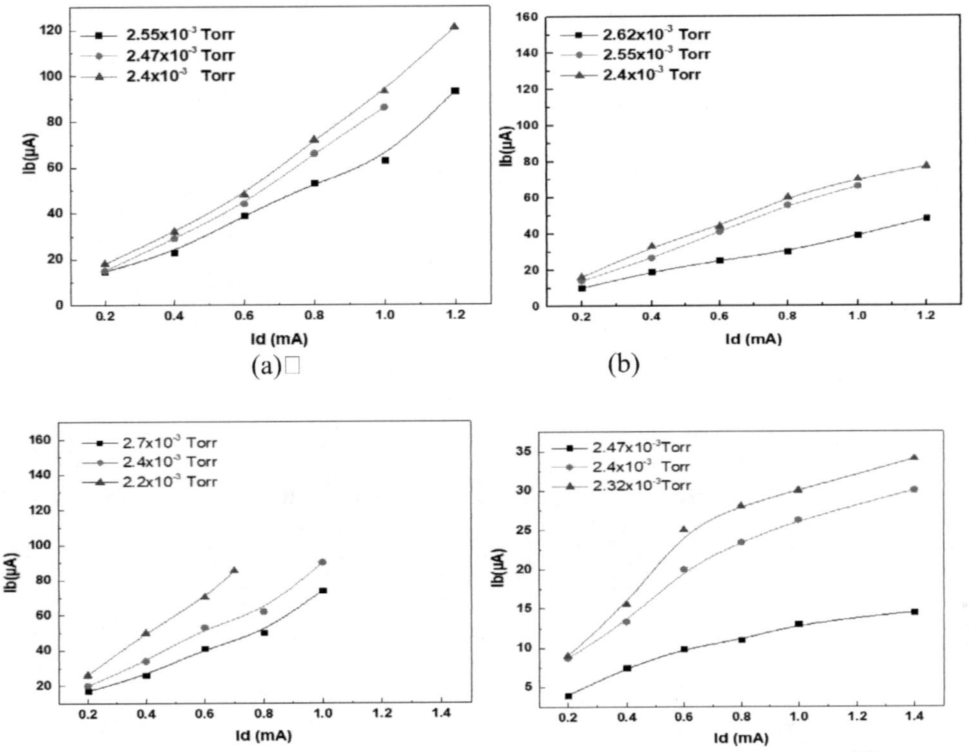

(a) (b)

(c) (d)

Fig. 3 Influence of the discharge current on the output ion beam current at changeable Ar^+ gas pressures and various anode and cathode distances: (a) 1 mm, (b) 2 mm, (c) 3 mm, (d) 4 mm.

Fig. 4 (a-f) shows the output ion beam characteristics at various Ar^+ gas pressures and increasing distances between the ion source and collector plate: (a) 1.5 cm, (b) 2 cm, (c) 3 cm, (d) 4.4 cm, (e) 5 cm, and (f) 5.3 cm. As seen in this figure, the discharge current is proportional to the output ion beam current [5]. Also, the optimum length between both the ion source aperture and the collector plate where found to be at 4.4cm where a maximum ion beam current output being 70 μA at P_{Ar^+} = $2.2x10^{-3}$ Torr and 0.6 mA discharge current due to the highest acceleration voltage and lowest argon gas pressure, also the beam waist was found to be narrow, around this point the ion beam blows up (Figs. 5&6).

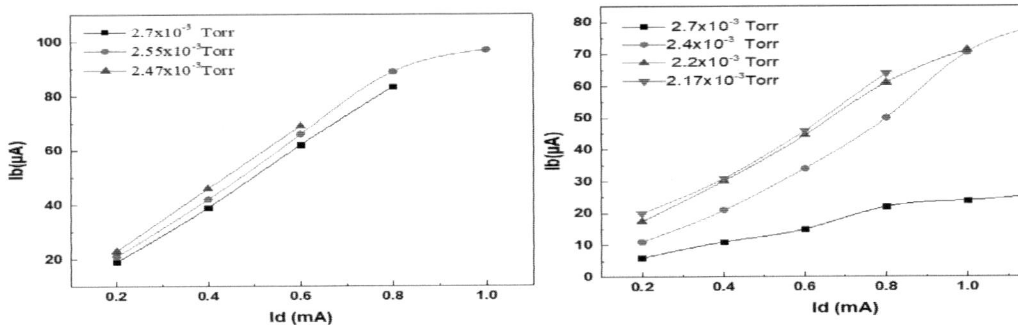

(a) (b)

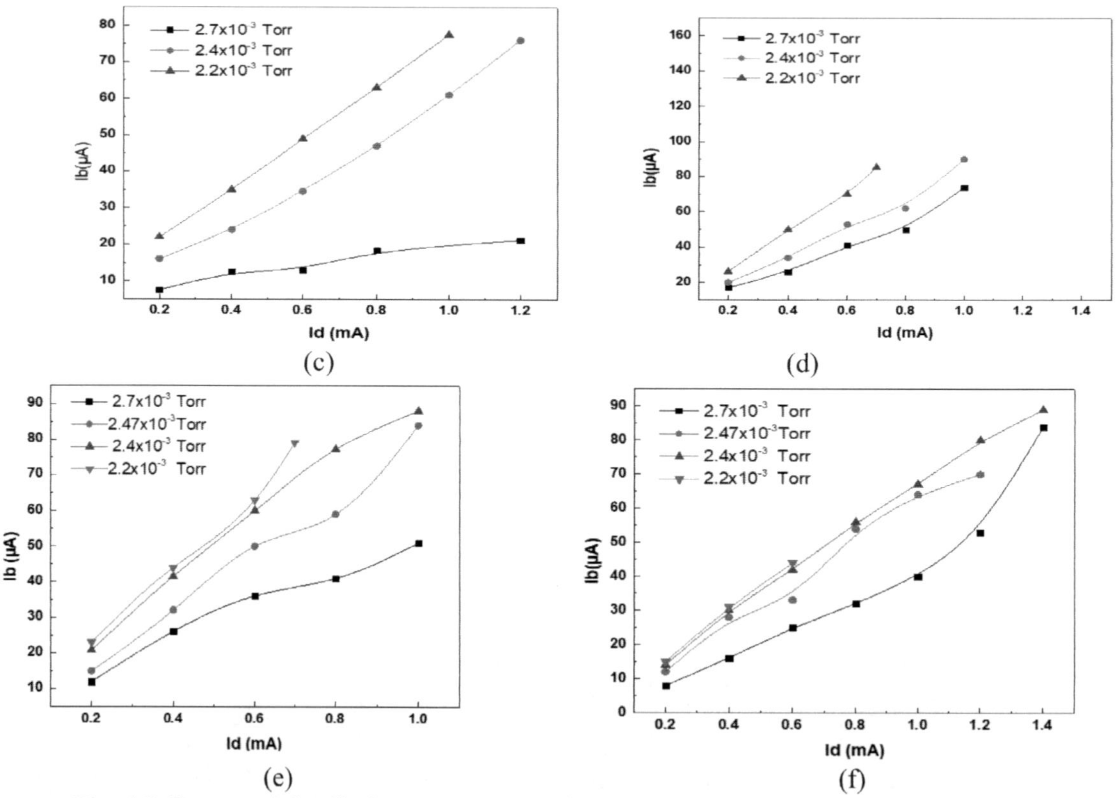

Fig. 4 Influence of the discharge current on the output ion beam current at different Ar^+ gas pressures and different distances between the ion source and collector plate: (a) 1.5 cm, (b) 2 cm, (c) 3 cm, (d) 4.4 cm, (e) 5 cm, (f) 5.3 cm.

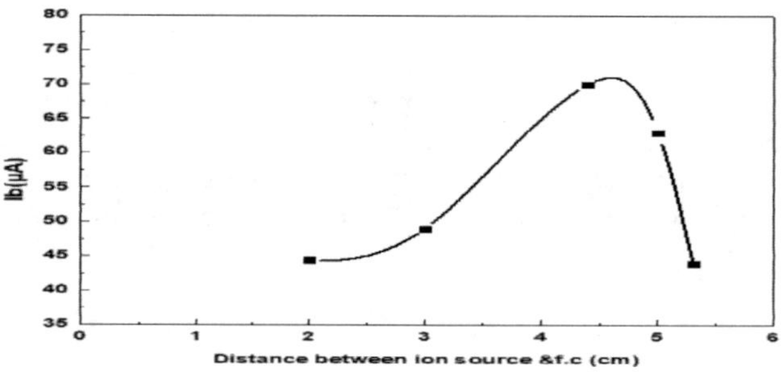

Fig. 5 Influence of distance between both the ion source and the collector plate upon the ion beam current at $P_{Ar^+} = 2.2x10^{-3}$ Torr and $I_d = 0.6$ mA

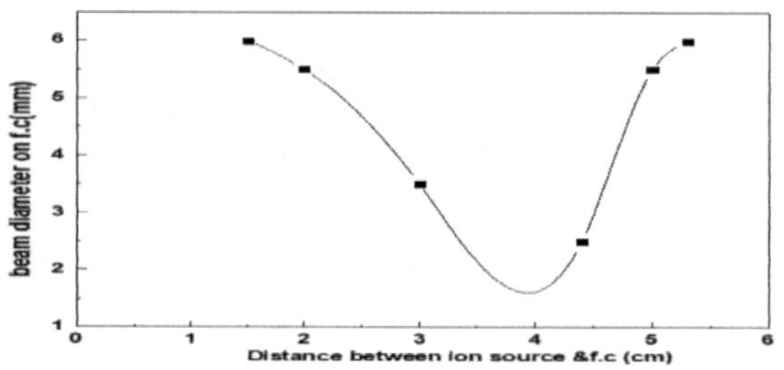

Fig. 6 Influence of distance between ion source and collector plate on beam diameter at $P_{Ar^+} = 2.2x10^{-3}$ Torr and $I_d = 0.6$ mA

Fig. 7 illustrates the relation between discharge pressure (Torr) and output current (μA) at I_d = 0.2 mA for four different gases: (a) argon, (b) oxygen, (c) nitrogen and (d) hydrogen. The rise of the discharge pressure reveals a significant rise of the output ions current due to an increase of gas molecules in the discharge bottle till a certain value. The rise in discharge pressure influences the plasma density to a point where colliding electrons obtain the maximum ionization cross section [6]. Intense pressure above this optimal range results in a reduction in plasma density attributed to a reduction in the mean free path and a decrease in the energy received by gas particles during collisions, resulting in a decrease in output current.

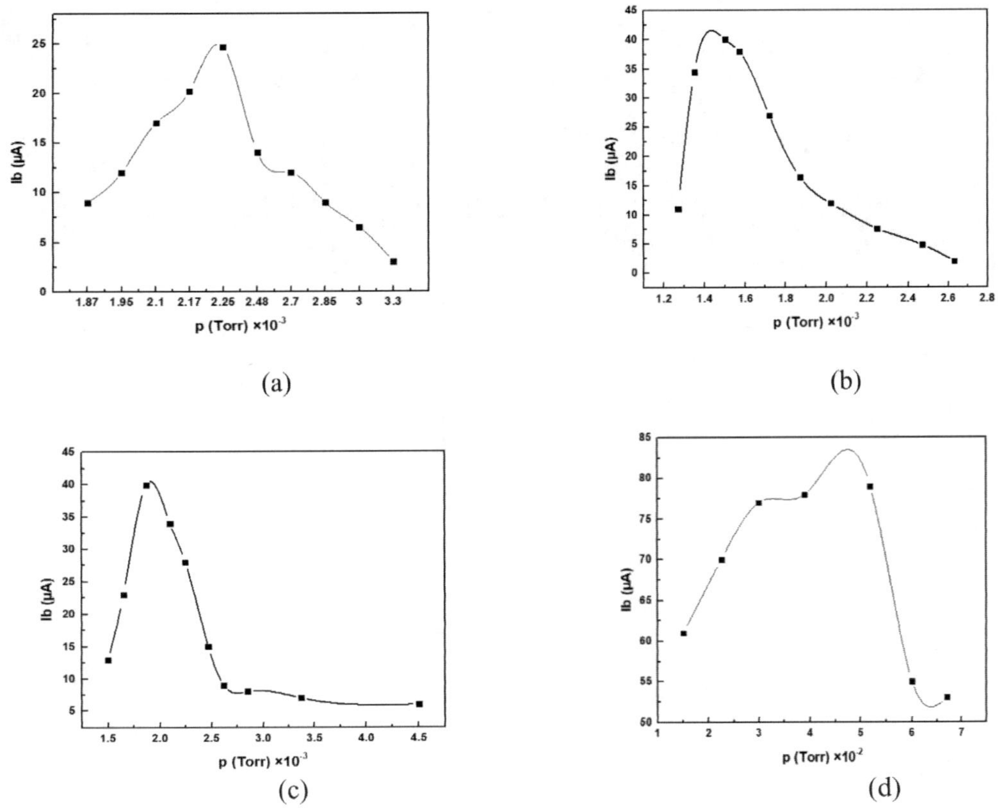

Fig.7 discharge pressure and comparable output ion beam current at I_d=.2 m for different gases: (a) argon, (b) oxygen, (c) nitrogen, (d) hydrogen

Fig. 8 shows that the output ion beam current for argon, oxygen, nitrogen and hydrogen gases are 70, 94,104 and 166 μA respectively, at the optimal operative condition of the newly developed home-made glow discharge ion source(the optimum length between both the cathode and the anode was found to be at 3mm & the optimum length between both the ion source aperture and the collector plate was found to be at 4.4cm), due to the difference in mass and ionization efficiency according to Child – Langmiur relation. The ion source has been operated and examined for different gases as tabulated below in table (1)

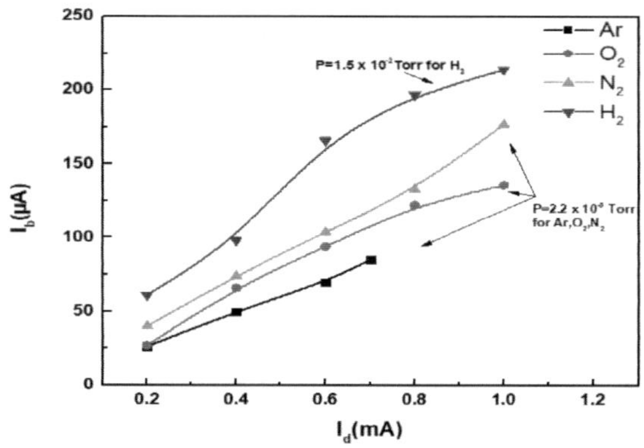

Fig .8 Influence of discharge current on ion beam current in various gases.

Table (1) Values of ion beam current for different gases

Beam energy range and current obtained at $I_d = 0.6$ mA		
Beam	**Energy (kev)**	**I_b (μA)**
Ar^+	4.6	70
O_2	4	94
N_2	3.8	104
H_2	3	166

Fig. 9 (a-c) demonstrates the connection between the extracted voltages V_{ex}, and the extracted argon ion beam current, I_{ex}, at various extractor electrode hole diameters (a)D1 =5 mm, (b) D2 = 7 mm, and (c) D3 =10 mm, respectively. It is obvious that as the extracted voltage increases, so does the extracted ion beam current [7]. Using these parameters, It was found that 7mm is the optimum extractor electrode hole diameter for attaining a maximum argon output ion beam current of (187μA) at $P_{Ar^+} = 2.2x10^{-3}$ Torr ,I_d= 0.6 mA discharge current and Vex.=4 Kv , as shown in (Figs. 9b&10) . It was also discovered that the maximum beam diameter was found to be at the extractor hole diameter of 7mm, as depicted in Fig. 11, where the electric field is optimal for focusing the beam at this hole diameter

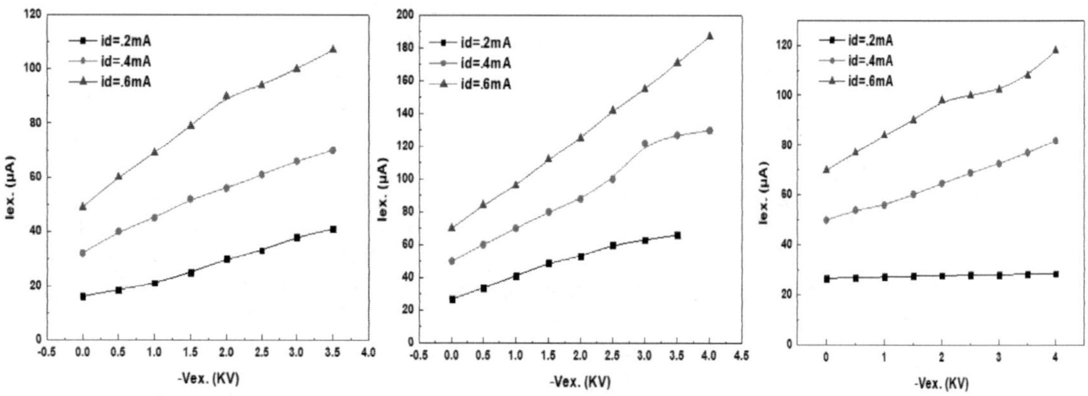

(a) (b) (c)

Fig. 9 The extracted voltage's influence on the extracted Ar+ current, at different extractor electrode hole diameters (a)D1 =5 mm, (b) D2 = 7 mm, and (c) D3 =10 mm

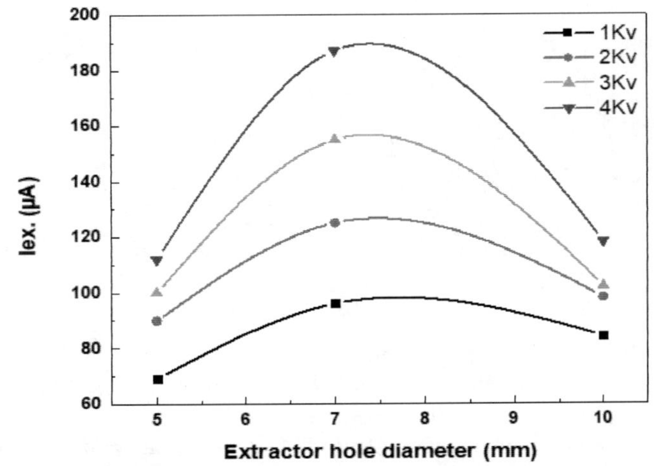

Fig. 10 Effect of the extractor hole diameter on the extracted Ar+ current at different extraction voltages, I_d=.6mA and p_{Ar^+} = 2.2x10^{-3} Torr

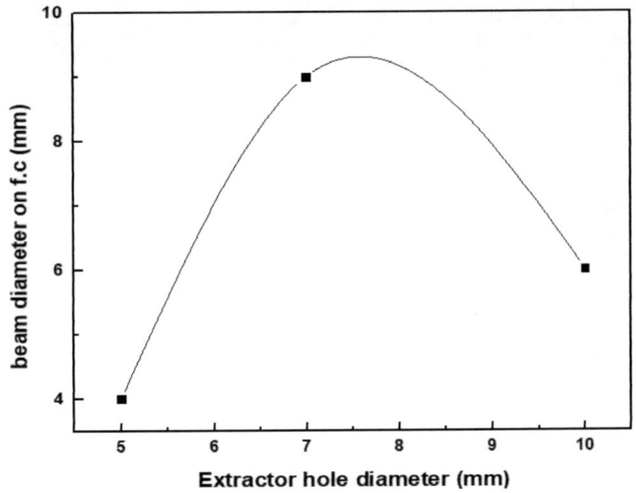

Fig. 11 influence of extractor hole diameter on beam diameter at different extraction voltages, I_d=.6mA and p_{Ar^+} = 2.2x10^{-3} Torr

Fig. 12 demonstrates the relationship between extraction voltage and the beam diameter at different distances between the ion source and collector plate. It was noticed that, the beam diameters at distances (30, 44, 50 and 53) mm respectively, reach their maximum values in the range of extraction voltages (700 v – 1300 v). on the other hand, at shorter distances (10, 15 and 20 mm), the beam diameters have a minimum value at extraction voltage 100 v and maximum values at extraction voltage range (300 v – 500 v) due to the interaction of electric field with the ion beam trajectories .

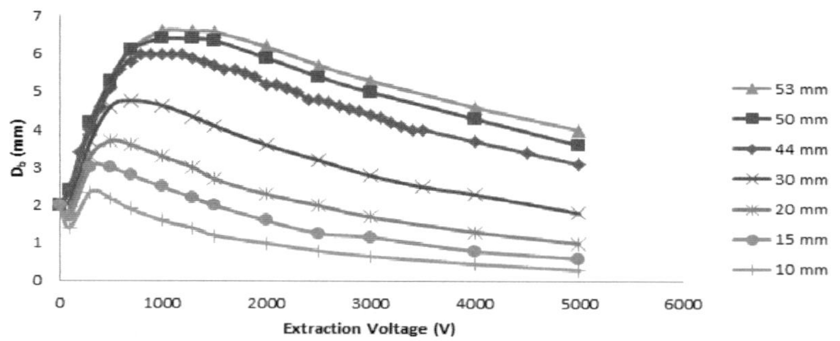

Fig.12 Impact of extraction voltage on beam diameter at different collector plate lengths from ion source aperture using SIMION computer program.

Fig. 13 depicts the relationship between the extractor holes and beam diameter, in which the beam diameter is proportional to the extraction voltages and the hole diameter as well, [8] as a consequence of reducing ion losses at the electrodes' edges, on the other hand, the beam diameter is inversely proportional to extraction voltage. The simulation and experimental measurements of an efficient ion beam extracted from the newly source agreed well.

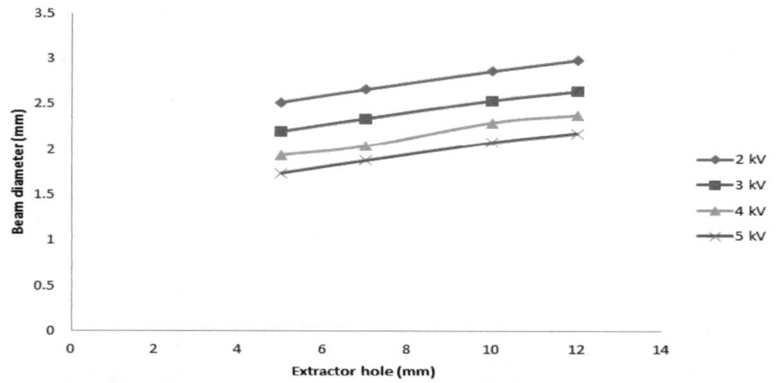

Fig. 13 Relation between the extractor holes and beam diameter.

Conclusion

A glow discharge ion source was designed, constructed and developed for extraction of low-power, high plasma density and high ion beam current. The designed and constructed home-made ion source has been successfully operated and examined. It was clarified that ,the optimum length between both the cathode and the anode was found to be at 3mm at which the source was able to deliver Ar^+ ion beam current up to 55 μA at 0.6 mA discharge current, and $2.4x10^{-3}$ Torr discharge .Beyond this optimal value, plasma instability and mechanical design issues manifest themselves, resulting in a breakdown of the system. Besides, it was discovered that the minor possible beam diameter was found at this specific distance (3 mm) due to the lowest pressure and the highest applied potential energy at this particular distance. Ideally, the plasma meniscus will have a concave shape under these conditions, producing a focused ion beam (minimum beam diameter).

Furthermore, it was observed that, the optimum length between both the ion source aperture and the collector plate was found to be at 4.4cm where a maximum output Ar+current is 70 μA at p_{Ar+}=2.2x10^{-3} Torr, 0.6 mA discharge current , due to the highest acceleration voltage and lowest argon gas pressure.

Moreover, the characteristics of ion beam current at the optimal operative conditions for different gases were obtained. The output ion beam current of argon, oxygen, nitrogen and hydrogen gases are 70, 94,104 and 166 μA respectively at 0.6 mA discharge current.

Additionally, It was found that, 7mm is the optimum extractor electrode hole diameter where a maximum argon current is 187 μA at p_{Ar+}= 2.2x10^{-3} Torr , I_d = 0.6 mA and $V_{ex.}$=4 Kv, In addition, the electric field was determined to be ideal at this extractor hole diameter, yielding a maximum beam diameter .

Finally, Ion beam optics has also been examined and studied using the SIMION computer program. It was noticed that, the simulation and experimental results of an efficient ion beam extracted from the ion source were found to be in good agreement.

References

[1] M.M.Abdelrahman,F.W.Abdelsalam,2010," Optimization of cold cathode saddle field ion source', Nuclear Instruments and Methods in Physics Research Section A: Accelerators, Spectrometers, Detectors and Associated Equipment, Vol. 613, Issue 2, pp. 185-189, https://doi.org/10.1016/j.nima.2009.11.044

[2] S. M. Mahmoud, A. M. Abdel Reheem, H. M. El-Sayed , H. El-Khabeary,2021, "Optimization and Characterization of Modified Cold Grid Cathode Penning Ion Source", Plasma Chemistry and Plasma Processing, Springer,vol. 41, pp.1535–1546.

[3] M. M. Abdel Rahman ,F. W. Abdelsalam,2012," Design and fabrication of a spherical configuration ion source", Europhysics Letters, Vol. 98, Number 4, pp. 5002

[4] F.W.Abdelsalam,A.G.Helal,M.M.Abdelrahman,B.A.Soliman,2009,"Operating characteristics of the hollow anode glow discharge ion source", ELSEVIER,vaccum, Vol. 84, Issue 3, pp. 405-409 , https://doi.org/10.1016/j.vacuum.2009.09.003

[5] Barakat A. Soliman, Moustafa M. Abdelrahman, Fatama W. Abdelsala, Kamal A.Aly,2013," Irradiation effect on PET surface using low energy argon ion beam", Journal of Nuclear Materials, Vol. 432, Issues 1–3, PP. 444-449, https://doi.org/10.1016/j.jnucmat.2012.08.006

[6] M. E. Abdelaziz, S. G. Zakhary, A. A. Ghanem, A. M. Abdelghaffar,1996," Preliminary results of a broad beam radio frequency ion source with electron plasma interaction", Review of Scientific Instruments ,vol.67, Issue 3, pp.1365 , https://doi.org/10.1063/1.1146668

[7] A. M. Abdel Reheem, M. M. Ahmed, M. M. Abdelhamid, and A. H. Ashour,2016," Verification of high efficient broad beam cold cathode ion source", Review of Scientific Instruments ,Vol.87,Issue 8, pp. 3302 ; https://doi.org/10.1063/1.4960394

[8] B. A. Soliman, M. M. Abdelrahman, A. G. Helal and F. W. Abdelsalam,2011," Simulation of ion beam extraction and focusing system", Chinese Physics C, Vol. 35, Number 1,PP. 83.

AUTHOR INDEX